A-Z
NANOCOMPUTERS

A-Z NANOCOMPUTERS

Dr. Albert Shawn

CENTRUM PRESS
NEW DELHI-110002 (INDIA)

CENTRUM PRESS
H.O.: 4360/4, Ansari Road, Daryaganj,
New Delhi-110 002 (India)
Ph.: 23278000, 23261597

B.O.: No. 1015, Ist Main Road, BSK IIIrd Stage
IIIrd Phase, IIIrd Block,
Bangalore - 560 085 (India)
Tel.: 080-41723429
Visit us at: www.centrumpress.com

A-Z Nanocomputers

First Edition, 2009

ISBN 978-93-80106-93-9

PRINTED IN INDIA

Printed at Mehra Offset Press, Delhi

Contents

Preface

Nanocomputer is the logical name for a computer smaller than the microcomputer, which is smaller than the minicomputer. (The minicomputer is called "mini" because it was a lot smaller than the original (mainframe) computers.) More technically, it is a computer whose fundamental parts are no bigger than a few nanometers.

This book intends to provide general, well-explained and practical knowledge regarding Nano computer In addition, it also makes clear that by gaining such knowledge what benefits he will derive. This book has two main purposes. One is to make the new teaching methods developed from these concepts and models more useful to teachers already somewhat familiar with them. The other is to inform other teachers of the possibilities for improved student understanding that these methods offer.

Author

Chapter 1

Nanocomputer

Nanocomputer is the logical name for a computer smaller than the microcomputer, which is smaller than the minicomputer. (The minicomputer is called "mini" because it was a lot smaller than the original (mainframe) computers.) More technically, it is a computer whose fundamental parts are no bigger than a few nanometers. For comparison, the smallest part of current state-of-the-art microprocessors measures 45 nm as of February 21, 2007. No commercially available computers that are named nanocomputers exist at this date, but the term is used in science and science fiction.

There are several ways nanocomputers might be built, using mechanical, electronic, biochemical, or quantum technology. It is unlikely that nanocomputers will be made out of semiconductor transistors (Microelectronic components that are at the core of all modern electronic devices), as they seem to perform significantly less well when shrunk to sizes under 100 nanometers..

NANO-COMPUTING

The history of computer technology has involved a sequence of changes from gears to relays to valves to transistors to integrated circuits and so on. Today's techniques can fit logic gates and wires a fraction of a micron wide onto a silicon chip. Soon the parts will become smaller and smaller until they are made up of only a handful of atoms. At this point the laws of classical physics break down and the rules of quantum mechanics take over, so the new quantum technology must replace and/or supplement what we presently have. It will support an entirely new kind of computation with new algorithms based on quantum principles.

Presently our digital computers rely on bits, which, when charged, represent on, true, or 1. When not charged they become off, false, or 0. A register of 3 bits can represent at a given moment in time one of eight numbers (000,001,010,...,111). In the quantum state, an atom (one bit) can be in two places at once according to the laws of quantum physics, so 3 atoms (quantum bits or qubits) can represent all eight numbers at any given time.

So for x number of qubits, there can be 2^x numbers stored. (I will not go into the logic of all this or this paper would turn into a book!). Parallel processing can take place on the 2^x input numbers, performing the same task that a classical computer would have to repeat 2^x times or use 2^x processors working in parallel.

In other words a quantum computer offers an enormous gain in the use of computational resources such as time and memory. This becomes mind boggling when you think of what 32 qubits can accomplish. This all sounds like another purely technological process. Classical computers can do the same computations as quantum computers, only needing more time and more memory. The catch is that they need exponentially more time and memory to match the power of a quantum computer. An exponential increase is really fast, and available time and memory run out very quickly.

Quantum computers can be programed in a qualitatively new way using new algorithms. For example, we can construct new algorithms for solving problems, some of which can turn difficult mathematical problems, such as factorization, into easy ones. The difficulty of factorization of large numbers is the basis for the security of many common methods of encryption. RSA, the most popular public key cryptosystem used to protect electronic bank accounts gets its security from the difficulty of factoring very large numbers. This was one of the first potential uses for a quantum computer.

"Experimental and theoretical research in quantum computation is accelerating world-wide. New technologies for realising quantum computers are being proposed, and new types of quantum computation with various advantages over classical computation are continually being discovered and analysed and we believe some of them will bear technological fruit. From a fundamental standpoint, however, it does not matter how useful quantum computation turns out to be, nor does it matter whether we build the first quantum computer tomorrow, next year or centuries from now. The quantum theory of computation must in any case be an integral part of the world view of anyone who seeks a fundamental understanding of the quantum theory and the processing of information."

NANOCOMPUTER

A nanocomputer is a computer whose physical dimensions are microscopic. The field of nanocomputing is part of the emerging field of nanotechnology. Several types of nanocomputers have been suggested or proposed by researchers and futurists.

Electronic nanocomputers would operate in a manner similar to the way present-day microcomputers work. The main difference is one of physical scale. More and more transistor s are squeezed into silicon chips with each passing year; witness the evolution of integrated circuits (IC s) capable of ever-increasing storage capacity and processing power. The ultimate limit to the number of transistors per unit volume is imposed by the atomic structure of matter. Most engineers agree that technology has not yet come close to pushing this limit. In the electronic sense, the term nanocomputer is relative. By 1970s standards, today's ordinary microprocessors might be called nanodevices.

Chemical and biochemical nanocomputers would store and process information in terms of chemical structures and interactions. Biochemical nanocomputers already exist in nature; they are manifest in all living things. But these systems are largely uncontrollable by humans. We cannot, for example, programme a tree to calculate the digits of pi, or programme an antibody to fight a particular disease (although medical science has come close to this ideal in the formulation of vaccines, antibiotics, and antiviral medications). The development of a true chemical nanocomputer will likely proceed along lines similar to genetic engineering.

Engineers must figure out how to get individual atoms and molecules to perform controllable calculations and data storage tasks.

Mechanical nanocomputers would use tiny moving components called nanogears to encode information. Such a machine is reminiscent of Charles Babbage 's analytical engines of the 19th century.

For this reason, mechanical nanocomputer technology has sparked controversy; some researchers consider it unworkable. All the problems inherent in Babbage's apparatus, according to the naysayers, are magnified a millionfold in a mechanical nanocomputer. Nevertheless, some futurists are optimistic about the technology, and have even proposed the evolution of nanorobots that could operate, or be controlled by, mechanical nanocomputers.

A quantum nanocomputer would work by storing data in the form of atomic quantum states or spin. Technology of this kind is already under development in the form of single-electron memory (SEM) and quantum dots. The energy state of an electron within an atom, represented by the electron energy level or shell, can theoretically represent one, two, four, eight, or even 16 bits of data.

The main problem with this technology is instability. Instantaneous electron energy states are difficult to predict and even more difficult to control. An electron can easily fall to a lower energy state, emitting a photon ; conversely, a photon striking an atom can cause one of its electrons to jump to a higher energy state.

SHRINKING COMPUTER

Civilisation has advanced as people discovered new ways of exploiting various physical resources such as materials, forces and energies. In the twentieth century information was added to the list when the invention of computers allowed complex information processing to be performed outside human brains. The history of computer technology has involved a sequence of changes from one type of physical realisation to another — from gears to relays to valves to transistors to integrated circuits and so on. Today's advanced lithographic techniques can create chips with features only a fraction of micron wide (a micron is a micrometre, i.e. a millionth of a metre).

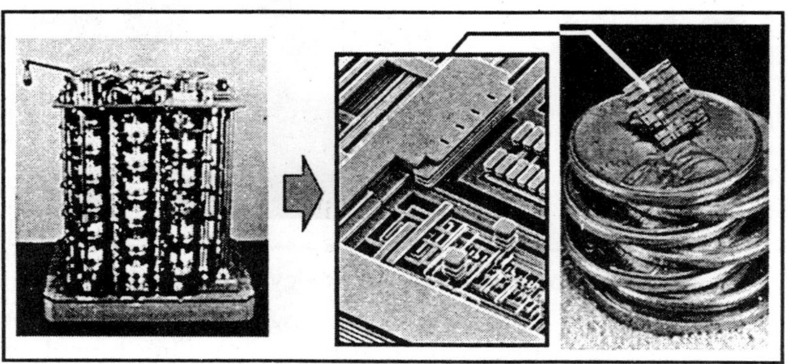

Fig. From the Beginning to the Present: on the left an Early Computing Machine Built from Mechanical Gears, on the right a State-of-the art IBM chip with 0.25 Micron Features. The Production Version will Contain 200 Million Transistors.

Nanocomputers

Over the last few decades computer power has grown at an amazing rate, doubling every couple of years. This increase is essentially due to the continual miniaturization of the computer's most elementary component, the transistor. As transistors became smaller more could be integrated into a single microchip, and so the computational power increased.

However this miniaturization process is now reaching a limit, a quantum threshold below which transistors will cease to function. Present 'state-of-the-art' components possess features only a few hundreds of nanometres across (a nanometre is a thousandth of a micron, or a billionth of a metre). If these chips were to be miniaturized further to the scale of tens of nanometres then their operation would be disrupted by the emergence of quantum phenomena, such as electrons tunneling through the barriers between wires. In order for the science of computation to progress further an alternative to transistor technology must be found, one whose components will function through quantum effects rather than in despite of them.

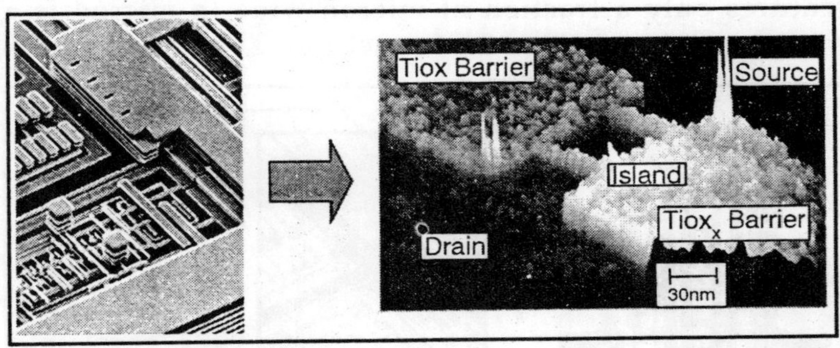

Fig. The Transition from Microtechnology to Nanotechnology.

The structure on the right is a single-electron transistor (SET) which was carved by the tip of a scanning tunneling microscope (STM).

According to classical physics, there is no way that electrons can get from the 'source' to the 'drain', because of the two barrier walls either side of the 'island'. But the structure is so small that quantum effects occur, and electrons can, under certain circumstances, *tunnel*.through the barriers (but only one electron at a time can do this!). Thus the SET wouldn't work without quantum mechanics.

As shown by the figure above, there are ways to redesign transistors to work using quantum effects. But it might be better to give up the idea of transistors all together, and use a completely new architecture that is more suitable for the nanometre scale. One such idea is shown below.

The figure shows how a particular circuit called a 'half-adder' can be made from a pattern composed of two kinds of cell. This type of architecture is very suitable for the nanometre-scale, where simple units form naturally. One way to make the cells would be using structures called quantum-dots, which are also know as 'artificial atoms'.

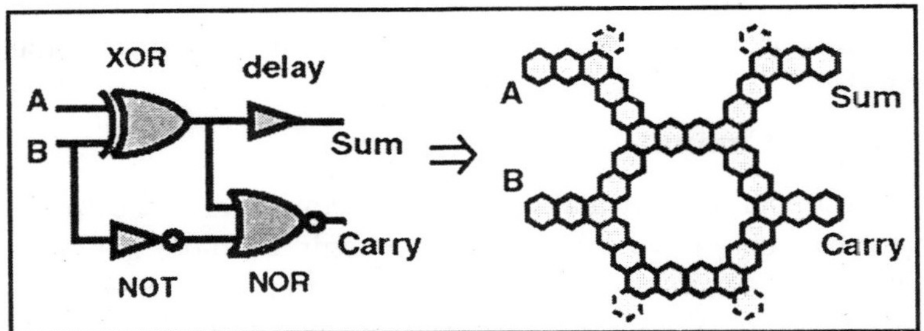

Fig. As an Alternative to using new kinds of Transistor, Nanocomputers might have an Entirely new type of Architecture made up of many simple units called 'Cells'.

The first generation of nanocomputers will have components that behave according to quantum mechanics, but the *algorithms* that they run will probably not involve quantum mechanics. We might call such computers 'nanometre-scale *classical* computers' (here the word 'classical' means 'not quantum'). But scientists have recently realized that there is another, more exciting possibility - quantum m.chanics might be used in an entirely new kind of algorithm that would be fundamentally more powerful than any classical scheme. A computer that could run such an algorithm would be a true 'quantum computer'.

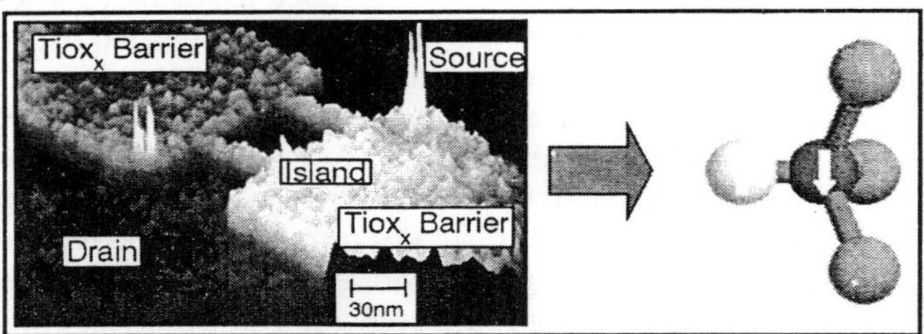

Fig. From an SET (on the left) to the Ultimate Computer Element: a Molecule! Although both these structures use Quantum Mechanics, only the one on the right could be employed in a true 'Quantum Computer'.

The 1H and 13C nuclei in isotopically labeled chloroform behave like small magnets, and interact with an external magnetic field. Nuclear spins can store and process information in the so called quantum superpositions (see below).

Quantum Computers

To explain what makes quantum computers so different from their classical counterparts we begin by having a closer look at a basic chunk of information namely one bit. From a physical point of view a bit is a physical system which can be prepared in one of the two different states representing two logical values — no or yes, false or true, or simply 0 or 1. For example, in today's digital computers, the voltage between the plates in a capacitor represents a bit of information: a charged capacitor denotes bit value 1 and an uncharged capacitor bit value 0.

One bit of information can be also encoded using two different polarisations of light or two different electronic states of an atom. However, if we choose an atom as a physical bit then quantum mechanics tells us that apart from the two distinct electronic states the atom can be also prepared in a coherent superposition.of the two states. This means that the atom is both. in state 0 and. state 1. There is no equivalent of this superposition in the classical world, it is a purely quantum mechanical phenomenon. Since we are used to seeing classical physics at work in the every day world, such quantum phenomena often seem counter-intuitive.

Now we push the idea of superposition of numbers a bit further. Consider a register composed of three physical bits. Any classical register of that type can store in a given moment of time only one out of eight different numbers i.e. the register can be in only one out of eight possible configurations such as 000, 001, 010,... 111. A quantum register composed of three qubits can store in a given moment of time all eight numbers in a quantum superposition.

This is quite remarkable that all eight numbers are physically present in the register but it should be no more surprising than a qubit being both in state 0 and 1 at the same time. If we keep adding qubits to the register we increase its storage capacity exponentially i.e. three qubits can store 8 different numbers at once, four qubits can store 16 different numbers at once, and so on; in general L qubits can store 2^L numbers at once (here 2^L means 2 to the power of L). Once the register is prepared in a superposition of different numbers we can perform operations on all of them. For example, if qubits are atoms then suitably tuned laser pulses affect atomic electronic states and evolve initial superpositions of encoded numbers into different superpositions.

During such evolution each number in the superposition is affected and as the result we generate a massive parallel computation albeit in one piece of quantum hardware. This means that a quantum computer can in only one computational step perform the same mathematical operation on 2^L different input numbers encoded in coherent superpositions of L qubits. In order to accomplish the same task any classical computer has to repeat the same computation 2^L times or one has to use 2^L different processors working in parallel. In other words a quantum computer offers an enormous gain in the use of computational resources such as time and memory.

Future registers in quantum computers may look like this ion trap.

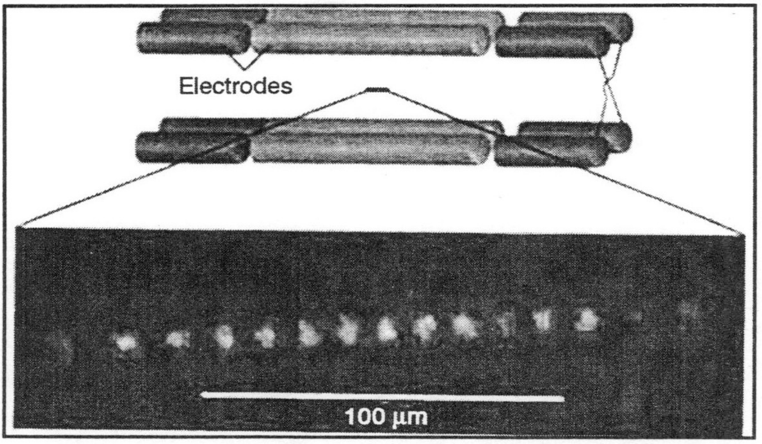

All ions in the trap have the same charge and repel each

other. Any motion of one of the ions is transferred by this electrostatic repulsion to other ions in the trap, inducing various collective motions known as phonons. A single ion can be set in motion by directing a laser pulse at that particular ion — each ion can be addressed separately because the inter-ion separation is much larger then the wavelength that induces the excitation. Combination of laser light and phonons can induce non-trivial logic and can be used in quantum computation. This kind of quantum logic gates are currently being implemented by experimental groups both in Europe and in the U.S.A.

Chapter 2

Compilation for Future Nanocomputer Architectures

Traditional compilation for high performance computers relies on a wide variety of sophisticated analyses and optimizations to translate a source language programme into efficient binary machine code for a specific target architecture. To perform this translation, the compiler parses the source programme, applying machine-independent optimizations on an abstract intermediate form of the programme.

Next, machine-dependent optimizations are performed with the goal of producing hopefully a significant improvement to performance of the resulting executable programme. These improvements are made with the specific attributes of the target machine, or processor, on which the programme will run in mind.

Different processors can differ dramatically in the code improvements which work best for them. For instance, a programme compiled to run optimally on one specific architecture may perform poorly on a different architecture. As a result, at least a portion of the compiler, the back-end, must be retargeted, potentially for each different processor or class of processors.

Although machine-independent optimizations can produce significant improvements in general, machine-specific knowledge, such as knowing the number of registers, memory organization and cache structure, instruction set, functional units available, and any other explicit parallelism that may be available, as with EPIC (Explicitly Parallel Instruction

Computing), Superscalar and VLIW (Very Long Instruction Word) processors, is required to achieve significant results on modern, high-performance architectures.

If Moore's Law is to continue to be applicable as a predictor, significant advances in chip manufacturing techniques are needed. In modern processors, feature sizes have been reduced while heat dissipation needs have increased to the point where there is a growing consensus that Moore's Law may actually be reaching an insurmountable barrier dictated by physics this time.

Nanotechnology, manufacturing performed through manipulation of atoms and molecules, is capable of overcoming this most recent barrier. Nanocomputer architectures, produced using this molecular manufacturing approach, provide a natural successor to current general-purpose microprocessor architectures.

Nanocomputers, of course, must be functionally at least as capable as their predecessors, fast, inexpensive, robust and capable of operating at room temperature and of executing legacy code. Whether Moore's Law truly loses applicability, or we merely readjust our thinking about it, it is clear that there are significant hurdles to overcome in computer design in the very near future. Nanocomputing is discussed in this paper within the context of the topic of reconfigurable computing, which includes Field Programmable Gate Arrays (FPGA), Field-programmable Custom Computing Machines (FCCM), cellular array architectures, and synthetic neural systems, among many others.

The attractive capabilities of these emerging technologies include the ability to dynamically reconfigure or redesign the functionality of the processor, to produce nanoscale features (transistors, gates, logic circuits) that are at the very limits of physics leading to processors which are many times smaller, more powerful and more efficient that those possible through current manufacturing techniques, and to generate a new, customized processor as simply as we now generate a new executable programme. In this paper we propose a compilation framework that could be used to produce a processor based on a source code programme such that this new processor would be ideally suited to running the compiled programme.

Much as traditional compilers customize the programme to suit the machine, we propose a compiler that customizes the machine to fit the programme. Our research is on the demand side of nanocomputing rather than the supply side; although no such reconfigurable nanocomputer architecture of such a grand scale yet exists, trends point to the development of molecular- and atomic-scale switches in the near future. We present the results of preliminary experiments designed to confirm earlier results in design space exploration within a high performance EPIC machine, and extrapolate these results to the increasingly realizable promise of nanotechnology.

BACKGROUND & MOTIVATION

With limitations of miniaturization within chip manufacturing approaching the bounds of what is physically possible with known techniques, new methods of manufacturing are needed. Although some success has been achieved with extremely specialized multiprocessor systems using current technology, such as IBM's "Blue Gene" project, such large scale multiprocessor systems suffer from significant interprocessor bottlenecks. Development of scalable interprocessor communication schemes for million or billion processor systems has been slow, although it is viewed as a fundamental challenge for the future of processor design. A wavefront approach to reconfiguration and interprocessor communication for such large scale processors has been proposed as a feasible approach.

Nanotechnology provides promise as a manufacturing approach that can address these current physical limitations, and indeed research in applications of nanotechnology to logic gate design and chip manufacturing is active and well-funded in the United States, Japan and elsewhere. There is little doubt that nanocomputers, consisting perhaps of millions or billions of reconfigurable processing cells, laid out in a vast, reconfigurable fabric, are technically feasible and likely to be realized. Our work attempts to design a compilation approach that will exploit the capabilities of these highly flexible future machines.

Nanotechnology

The field of nanotechnology is a promising approach to manufacturing through the direct manipulation of atoms and molecules, accomplished through some combination of chemical, electrical and physical interactions. Although the hype surrounding the great promise of this potentially revolutionary technique is vast, reminiscent of the ballyhoo in the 1970s regarding the coming age of artificial intelligence (a robot in every home to do one's bidding), it is undeniable that it is a viable manufacturing approach that is gaining considerable momentum. Nanotechnology consists of a group of technologies, just emerging, which enable matter to be manipulated at the nanometer scale. This infinitesimal scale, on the order of a small number of atoms, is at the lower limits of physics as they are now understood, thus their applicability to solving current feature-size hurdles of chip manufacture, for example.

Less precise applications of nanotechnology are already producing useful products in the pharmaceuticals industry, and the techniques are emerging that will allow more sophisticated products to be produced using much more precise control of matter.

Although there is controversy surrounding the future benefits of nanotechnology, current trends point to the development of artificial molecular machine systems capable of building complex systems to atomic precision. In effect, we will duplicate the ability of, for example, a cow to convert hay into beef, a feat of molecular engineering that is currently performed quite effectively by the cow. The result of this bottom-up approach to manufacturing could be products that cure cancer by repairing cells, replace fossil fuels with clean and efficient alternatives, change the colour of paint on a wall with the touch of a finger, and generate massively parallel and reconfigurable computers the size of a thumbtack.

Microwave oven sized advanced molecular manufacturing systems will be capable of making large, useful products, very cheaply and with tremendous precision. Acceptance of these provocative predictions is contingent, of course, upon demonstrable results. Along with the predictions of amazing innovations come those of enormous danger. An example of such a nanotechnological nightmare is the "gray goo" problem, where a self-replicating nanomachine, and its exponentially increasing progeny, run amuck, breaking down all of the matter encountered into a uniformly gooey gray substance. The objective, then, is to proceed cautiously with the realization that a "bug" in a nanotechnology programme could potentially cause more damage than one that now leads to a seemingly quaint Blue Screen of Death. Exhaustive research is critical to the establishment of safe approaches to the development and use of nanotechnology.

Reconfigurable Architectures

As mentioned before, a wide variety of successful hardware approaches exist for reconfigurable computing. As physical limitations of feature size reduction and heat dissipation are reached, nanotechnology provides an approach to overcome these limitation. Because nanotechnology could lead to inexpensive production of highly reconfigurable computer hardware, it is natural to extrapolate the current state of the art to this emerging technology. Research strongly suggests that reconfigurable architectures, if efficient, will provide a better fit and thus improved performance for general purpose computation. Recent advances in design space exploration as applied to compilation for FPGA-based and other reconfigurable architectures demonstrates the performance improvements possible with customized architectures.

The concept of "programme in, chip out" (PICO) relies on compiler analysis, particularly targeted automatic parallelism, to identify programme fragments that will benefit most from customized hardware. Future nanocomputer architectures will be formed from non-volatile reconfigurable, locally-connected hardware meshes that merge processing and memory. This architecture models more closely than any existing computer architecture that of the human brain, with its vastly superior computing power. Although the nanocomputer does not yet exist, the performance of any programme can benefit from reconfiguration of the machine on which it runs to better match the resource requirements of the programme. In contrast to an approach using design space exploration, the potential flexibility of nanocomputer architectures means that configuration will not be limited to a known design space.

Architectural Features & Compiler Analysis Techniques

Certain characteristics are common to most processors, and should be considered as candidates for reconfiguration. Characteristics such as the number of registers, functional units, memory units, cache and memory organization, branch history and prediction support, pipeline organization, and the instruction set architecture (ISA) should be considered. Compilers currently retarget a source programme to a target machine. With the flexibility of machine reconfiguration, the task of retargeting is made more powerful because the resulting target machine will be better suited to the specific characteristics and requirements of the source programme. In analyzing the characteristics inherent in a given source programme, the ability to detect instruction-level, tasklevel and coarse-grained parallelism will also guide the description of architectural features.

For instance, the amount of instruction-level parallelism (ILP) available in a programme can impact the number of functional units that are specified, leading to improved performance. Rather than computer architects investing significant money and time in the design of an excellent general-purpose processor that does reasonably well running most source programs, the compiler itself can generate a well-tuned machine configured to match the individual source programme, allowing each to maximize performance. The compiler has the responsibility of analyzing the source programme and generating a suitable machine description based on the results of the analysis. There are extensive analyses available, although the majority is designed to optimize the programme for a specific target machine.

Any analyses that measure or enable the measurement of variable liveness, inherent parallelism, branching behaviour, common subexpressions, memory access patterns, cache utilization, register allocation opportunities, instruction scheduling opportunities, and many other characteristic-related metrics are candidates. All can provide information to guide the generation of a machine description that is suited to the source programme. The key consideration is the ability to gather information about static characteristics or predictive behaviour of the source to enable production of a best-fit machine description.

Compiler Design

The major goal of the initial work reported in this paper is the design of a compiler that generates as its output both an executable version of the original source programme and the description of a machine on which the executable will run well. This study focuses on ILP architectures, although the concepts also apply more broadly to include coarse-grain approaches. Traditional compilers take the source code and translate it into a binary form suitable for a specific processor, optimized to run as well as possible on that target machine. Knowledge of the target machine is needed to perform machine-dependent optimizations. Our approach differs in that the configuration of the target machine is unknown when compilation begins. The machine configuration is extracted from the source programme, based on its resource requirements. In this way, the resulting machine is an excellent fit to the programme.

To discover this machine description, the compiler must perform a number of analyses designed to reveal the needs of the programme. For example, live variable analysis can be performed to help determine how many registers are needed. However, the number of registers needed also depends on the issue width of the target machine, which is determined by the inherent ILP in the programme. Thus, there are numerous interactions to consider when analyzing a programme and generating a machine description.

Figure illustrates the organization of the proposed nanocompiler from a high level. Source code is processed by the Front end of the compiler, including machine-independent optimizations.

The resulting intermediate form is passed to a Machine requirements analysis phase, which performs static analysis, providing metrics to the Machine description generation phase.

The resulting machine description is used by the Processor generator phase to generate or reconfigure the target machine, and by the compiler Back end to perform machine-dependent optimizations and generate the executable code. The organization in Figure also shows how information from the Processor generator and even runtime profiling information could be fed back into the Machine requirements analysis phase to enable iterative refinement of the machine description, and thus of the processor itself.

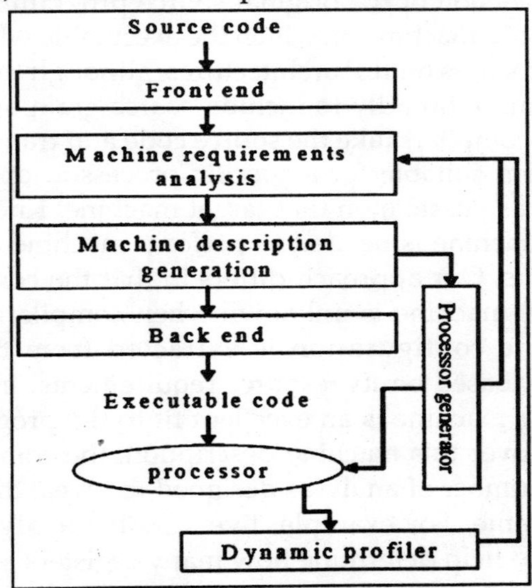

Fig. Organization of Nanocompiler

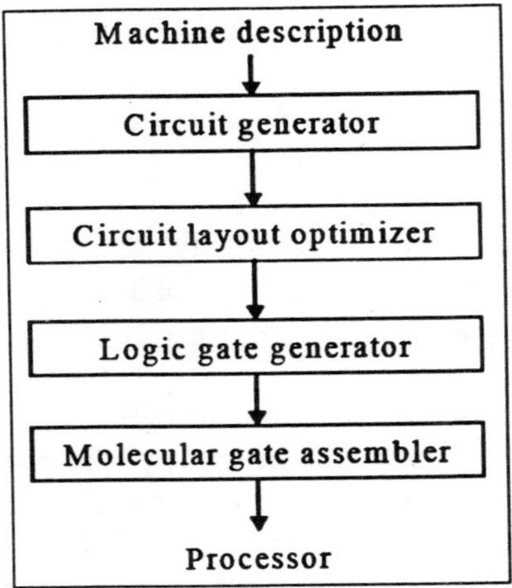

Fig. Organization of a Nanotechnology based processor Generator

If more of the functionality of the compiler is included in the operating system or hardware, dynamic reconfiguration of the processor is possible while a programme is running. Just-in-time load-time configuration involving an analysis of the requirements of a programme could be incorporated into the processor's loader, as well. When a programme is run on a nanocomputer with a JIT reconfiguration model, the machine could be reconfigured to better suit the programme, thereby improving its performance.

Inside the Processor generator, either a machine is reconfigured or it is generated using a nanotechnology approach.

Figure shows the organization of a nanotechnology-based processor generator. The machine description is analyzed through a sequence of phases that translate the description into a layout of circuits that implement the machine, which in turn is implemented using logic gates, which are assembled using molecular manufacturing techniques, to produce the target processor.

Experimental Study

The prototype of our compilation approach has been implemented as an extension of the Trimaran research compiler system. Trimaran was selected for its reconfigurable EPIC simulator and ease of extension. For this research, we modified Trimaran by isolating output from the live variable analysis and scheduling modules to incorporate their data into a machine description generator.

Experiments in this initial study were facilitated with the use of a number of high-level control scripts and drivers that were developed as part of related ILP-targeted compiler optimization research.

Experiments were designed and performed to compare the impact of our nanocompiler approach on programme performance. The experiments were conducted on the full set of eight C benchmarks in the SPEC95 CINT suite.

Programs were compiled and run on five pre-configured EPIC processors and a 6th custom generated EPIC processor. These six configurations consisted of 1-4 functional units (both integer and floating point), 1-4 memory units, 1-2 branch control units (to manage branch history, prediction, etc.), and registers proportional to the number of functional units (64 per).

Note that the 6th custom generated processor varied in number of functional, memory and branch units. These simulated processors produce copious static and dynamic statistics, including machine cycles consumed, operations performed, memory and cache access behaviour, and processor utilization.

The count of machine cycles was used to calculate speedup, since the simulations in Trimaran are performed independent of any specific clock cycle frequency. We initially expected that a carefully hand-configured general-purpose processor would provide the best overall performance for the entire suite of benchmarks. The results in Table indicate that reasonable and common machine configurations do indeed produce good performance speedup.

Table presents speedup for four machine configurations, as well as a custom, derived, machine configuration as compared with a 6th sequential, baseline processor configuration.

Table. Comparison of speedup for various architectural configurations.

Benchmark (lines of source)	Machine #2 2F,1M,1B Cost: 140	Machine #3 2F,2M,2B Cost: 260	Machine #4 4F,2M,2B Cost: 280	Machine #5 4F,4M,4B Cost: 520	Machine #6: Generated machine		
					Config	Cost	Speedup
099.go (29,713)	1.55	1.67	1.71	1.71	2F,1M,2B	160	1.69
124.m88ksim (20,026)	1.89	2.31	2.33	2.33	2F,2M,1B	240	2.31
129.compress (2,322)	1.95	2.23	2.25	2.74	2F,4M,1B	440	2.71
130.li (7,683)	1.30	1.39	1.39	1.40	2F,1M,1B	140	1.39
132.ijpeg (32,214)	1.91	2.11	2.22	2.32	4F,4M,1B	460	2.32
134.perl (27,768)	1.36	1.60	1.60	1.60	2F,2M,1B	240	1.60
142.gcc (211,484)	1.61	1.89	1.89	1.89	2F,2M,1B	240	1.89
147.vortex (67,286)	1.91	2.21	2.31	2.35	2F,4M,1B	440	2.35

For each configuration, numbers of functional units, memory units and branch control units are shown. For example, "4F, 2M, 1B" signifies a machine configured with 4 functional units, 2 memory units and a single branch unit. Also shown in Table is a cost for each configuration, based on a cost function C, where f, b and m are the numbers of functional, branch and memory units, respectively. The cost function is:

$$C = (fxF) + (bxB) + (mxM)$$

The values F, B and M are constant weight assigned to functional, branch and memory units to attempt to assign cost based on the complexity and increased latency of increases in the numbers of each type of unit. The values assigned to these are: F=10, B=20, M=100. For example, the calculated cost of Machine #1, the sequential processor (1F, 1M, 1B), is 130. Speedup either improved or remained steady as the cost of the simulated machine increased. Benchmarks that exhibited more memory intensive operations (129.compress, 132.ijpeg, 147.vortex) tended to continue to improve as the number of memory units increased.

For benchmarks that were more computationally intensive (124.m88ksim, 132.ijpeg) or had significant data or control dependencies that constrained ILP (099.go, 134.perl, 142.gcc), additional functional or branch units tended to improve performance more than memory units. The default configuration in the Trimaran system is Machine #3, which exhibits reasonably good speedup for all benchmarks. Results of custom generation of a machine as the result of compile-time analysis of each benchmark is shown in the last three columns of Table. Note that configurations tend to reflect a reduction in the number of units when additional units did not tend to improve performance.

Most notably, the cost of each generated machine is less than the cost for comparable speedup provided by one of the preconfigured machines, and the Machine requirements phase appears to have frequently identified a best, or at least better, fit machine configuration for each benchmark. Within the context of the EPIC architectures possible in Trimaran, it appears that our compiler framework leads to reduced cost, improved speedup, or a balance between the two when there are competing factors. In some cases, four memory units did not lead to speedup over a configuration with two memory units, and the machine requirements phases arrived at the correct conclusion.

Within the relatively constrained domain of EPIC, VLIW, Superscalar and similar ILP architectures, this approach may well be a feasible addition to current compilers that target reconfigurable versions of these machines. Because a hypothetical nanocomputer architecture is quite large (billions of processing units on a single chip), the machine description generation task is more difficult. With increased flexibility and freedom of reconfiguration comes increased responsibility on the part of the compiler phases to thoroughly analyse the source programme and produce a well-fitting machine description. There are improved opportunities for parallelism, and the potential to redesign the instruction set, pipeline organization, and any other feature of the machine.

Ultimately, the promise of this computational nanotechnology is straightforward, on-demand molecular-level manufacturing, which means that a new processor could be specified, manufactured, used and discarded (or recycled). Although such capabilities sound far-reaching and futuristic, from a practical standpoint such a technological revolution is technically and physically feasible, if far-off.

Related Work

Research in the area of FCCMs where a system combines reconfigurable logic with a general-purpose processor has demonstrated speedup . This research has focused on architectural details of the reconfigurable hardware and compiler features that exploit it. Our research attempts to build on this work by increasing the capability of the compiler to generate more of the target machine. Integration of a reconfigurable functional unit into a superscalar RISC processor improves upon earlier FCCM research by focusing on the hardware-compiler interface, although it relies on hand modification of source programs to achieve speedup. With our approach, all analysis and modification would be performed by the compiler.

An innovative approach to large-scale, homogeneous, undifferentiated, reconfigurable architecture improves upon FPGAs using a less expensive nanoscale "cell matrix" approach. This work describes how networks of atomicscale switches can be configured in parallel and used to fabricate scalable processors that are customized to specific tasks. Our research is targeted to a nanocomputer cell matrix architecture, focusing on use of the compiler to automatically generate reconfiguration instructions.

Hewlett Packard's research on PICO, a system that automatically designs custom computers is very similar in its goals. With PICO, parallel portions of a programme run on reconfigured FPGA hardware while the remainder runs on a connected, embedded processor. PICO employs design space exploration, selecting the best configuration from among a set of pre-designed configurations.

While PICO targets primarily embedded processors and uses design space exploration, our approach envisions a desktop or embedded computer that reconfigures its own hardware to any arbitrary configuration, using machine requirements analysis and nanotechnology. This process may resemble FPGA reconfigurability, or actual physical molecular reassembly, performed at compile-time or runtime.

The field of nanotechnology is quite active and rapidly advancing, with frequent accomplishments being made in many disciplines including medicine, pharmaceuticals, chemistry, physics, and of course, computer engineering. Little has been done in the area of compilers and optimization specifically aimed at using this technology. Our research attempts to pursue basic and applied nancompiler research and encourage others to do the same.

Summary and Future Work

Reconfigurable computing is an emerging area of hardware, and the promise of nanotechnology to hurdle looming barriers to continued applicability of Moore's Law is being recognized. We have designed a compiler framework targeted to a nanocomputer, and demonstrated that the techniques of machine description analysis and processor configuration-code generation using our compiler can improve performance while limiting costs.

Although our implementation was restricted to a reconfigurable EPIC architecture simulator, as it was the most appropriate and available platform at the time of our study, there are similarities in the requirements of all processors, including nanocomputers. With the significant flexibility and capability of nanocomputers, it seems that the responsibility for guiding the configuration will fall to the compiler. This research demonstrates the feasibility of that approach.

Where once the compiler customized the programme to suit the machine, the compiler will soon fully customize the machine to suit the programme, extending code generation to include code that will reconfigure, or even guide the manufacture of the processor itself.

We are planning extensive research and experimentation in the area of compiler optimization for nanocomputers. We are currently developing a compiler analysis framework that simulates generated architecture specifications using Colored Petri Nets. Additional work is being done in compiler approaches to generating and simulating nanomechanical computational machine descriptions from source code, applications to low-power and embedded computing, automatic parallelization techniques and description of nanocomputer architectures.

Chapter 3

Designing A Nanocomputer

CURRENT ARCHITECTURE PARADIGMS

The last twenty years of computer hardware development have relied on the silicon chip, namely the CMOS integrated circuit. CMOS stands for a complementary metal-oxide semiconductor, and its invention is the cornerstone of digital integrated circuits. These transistors have been shrinking in size and increasing in speed each year as famously predicted by Gordon Moore. Computer architects are now able to squeeze countless transistors on a single silicon chip. An article in PC Magazine state that "Intel estimates that it will offer chips that run at 20 GHz and contain 1 billion transistors by 2007."

While Moore's Law has held up for a significant amount of time, many foresee the end of the massive hardware improvements in the next decade. The rising chorus calling for the demise of Moore's Law is almost deafening with every new nanotechnology proposal put forth.

Below is an overview of the major components of today's computer, along with a brief description of their current design, and some conjecture as to what might become of these components in the future. Before we can propose a restructuring of a new nanocomputer, this overview will allow us to gauge where we are today so we can contrast it with the direction future computing trends are going.

CHIP MANUFACTURING TECHNIQUES

Current CMOS chips are manufactured using lithographic-fabrication, which is a top-down approach to integrating circuits. Most of the expansion which has been made over the last 20 years took place by scaling down the size of transistors so that they can fit on top of the silicon chip. The CMOS design is centered on using a pair of transistors for logic functions which complement one another.

These chips deal well with power usage and heat dissipation, and have been the central manufacturing technique on the market since the 1970s. Current sizes of the transistors on CMOS chips are reaching the nanometer size, and continued improvement has been proven to be limited.

Manufactures such as Intel see the end of the evolutionary path set out by Moore and have begun looking for alternatives to CMOS lithographic fabrication. While lithography has been a big part of why we have seen such incredible improvements in size and speed of chips in the last 20-30 years, many are seeing that this current technique is not scaling well, and predicting the end of CMOS advancement.

In the computing industry, chip manufacturers outline a roadmap for their manufacturing each year called the International Technology Roadmap for SemiConductors. In their most recent meetings, they have stated, "Eventually, toward the end of the Roadmap or beyond, scaling of MOSFETs (metal-oxide semiconductor field effect transistor) will become ineffective and/or very costly, and advanced non-CMOS solutions will need to be implemented."

While lithographic manufacturing of CMOS chips has been wildly successful, a new fabrication technique will be needed to continue the current pace of improvements in speed and size.

However, changing the computing industry's reliance on CMOS will be a major paradigm shift. While Keyes argued in 1985 that no other device has matched silicon with respect to gain, voltage restoration, noise reduction, tolerance to variability in manufacturing, isolation of inputs from outputs, excellent fan-out, ability to invert a signal, tolerance to the high demands of a system environment, as well as a huge body of research and financial commitment, the end of the reign of the silicon chip is being predicted time and again.

Some possible alternative CMOS solutions will be discussed in the next chapter.

INSTRUCTION SET DESIGN

Instruction Set Design has moved from the very simple accumulator, to the advanced use of registers and load–store architecture. The most recent design shift for instruction sets was the move from CISC (Complex Instruction Set Computer) to RISC (Reduced Instruction Set Computer). CISC was used such machines as the DEC Alpha and the HP3000. Intel's x86 processor is the only remaining computer in production which still uses CISC. Most computers today have RISC architecture.

The RISC philosophy of design came into dominance due to the increasing complexity of CPU designs and the need to speed up instruction executions. It is also well known that RISC chips require fewer transistors to execute their instruction. All RISC instructions can be executed in a single clock cycle, and more complex instructions are handled by stringing simple instructions together. In contrast, CISC has many more instructions, and relies less on the compiler to translate to assembly. RISC uses less RAM as well, and its simplicity was appropriate when hardware was very slow and very expensive.

There has been a fair amount of debate in recent years regarding innovation in instruction set architectures in order to increase performance. revious performance gains have relied on improvements in instruction level parallelism and thread level parallelism.

Most of the hesitation behind changing instruction sets has to do with the disruption it can cause to end users. The gains in speed and performance have been the driving force behind a design change. It's likely that only a complete paradigm shift in the design and manufacturing of computers will cause any change in the basic instruction set design of the modern computer in place today, much akin to the movement from CISC to RISC. It is highly likely that a move into Nanocomputing would be a catalyst for another shift in instruction set design, from RISC to perhaps an instruction set which relies less on software and takes advantage of significant gains in hardware speed and efficiency.

Multiprocessing and Instruction Level Parallelism

Multiprocessing came into being with the advent of the UNIX operating system, in which several computers could be networked together for computing power. CPUs can be linked in a tightly coupled fashion and connected by a bus, or loosely coupled by simply networking computers in a cluster.

Either fashion is limited by what is known as Amdahl's Law. The law is a formula to measure the increase in speed gained by adding another CPU to an algorithm's computation. There is a limit and an overhead associated with the addition of a new CPU, and the overall speedup increase will only be fraction of the true computing speed of the additional CPU. The use of multiprocessing does provide a significant benefit, but this benefit has a finite ceiling. Multiprocessing has been used in industry a great deal through the implementation of clusters.

Instruction level parallelism is the execution of instructions in parallel, by overlapping instructions which do not depend on one another in order to complete. While this provided a boost in processing speed initially, most CPUs simply do not have that many instructions which they can execute in parallel.

According to an article in Embedded Systems Programming, "ILP becomes its own worst enemy. Even if you could find two, four, or 50 instructions to execute simultaneously, you probably couldn't access all the data at once. Data starvation becomes a real problem.

For high-end systems, bandwidth is more important than ILP. The processor in your Macintosh, SPARC station, or PC spends more cycles waiting on memory than it does processing instructions.

(Depressing, huh?) Improving your "CPU efficiency" will simply increase the number of wait states." [TURLEY] Perhaps the promise of speed and efficiency by parallelism will never reach the heights we had hoped within the current architecture paradigm, until the bottleneck in memory and bandwidth is corrected. Nanotechnology has promise in the area of multiprocessing and clustering, because adding more CPU processors and linking them will increase computing power. However, true computing increases will occur with changes in buses and memory.

MEMORY

Cheap and fast memory has become more readily available rather recently, but memory more often than not is the place where backlog begins. Computer scientists have focused a great deal on faster caching techniques and more efficient memory management algorithms, all while the cost of the hardware continued to get smaller and go down. It's presumed that nanocomputing will allow memory to move out of this position of bottleneck because caches can become smaller, faster and simpler, however it's not clear at this point if that will be true. It's likely that computers will be faster and smaller all around, but the weakest links in any computer will still be the memory and interconnects.

The biggest improvement in main memory bottlenecks in modern computing has come from the use of virtual memory, in which a computer can act as though it has much more RAM than it actually does by using the hard disk to copy data. Virtual memory is used in all modern computers to run today's software, which tends to demand a great deal of space to run. Swapping algorithms and small caches have added to virtual memory by allowing programs to run quickly enough to pick up the slack for when a disk write is necessary.

Interconnections

Current PCs use buses for the interconnection between devices. The bus is a commonly overlooked component to a computer's size, speed and efficiency. The bus ties almost everything together, including the hard disk, the CPU, the memory, and any other I/O components a computer might have.

The PC generally has two main buses. The first, known as a system bus, connects the CPU and the memory. The second, which is a bit slower, is used to communicate with other devices such as the hard drive or sound/ video cards. Modern PCs also usually offer a USB which stands for "Universal Serial Bus" to allow users to connect any number of peripherals to the computer. These peripherals can include extra memory or extra hard drives. The system bus is an area in the computer which creates bottleneck between the hard drive and the CPU. As system buses have evolved, they have been able to deliver more data at a faster speed from the hard disk and into memory. This is an area in which research in nanotechnology can have a profound and marked effect.

Storage

Current technology trends have measured that disk capacity doubles approximately every 18 months. This increase is driven by the embedded computing market. Computers today can be found in everything from phones to music players to cars, and smaller disks are needed to run the software. The average computer has gone from having a 5 gigabyte hard drive in 1997 to an average 80 gigabyte hard drive in 2005. The costs of hard drives have decreased in almost the same proportions.

It is in this area that nanotechnology could also find its largest impact. As stated in the introduction, one of the main visions of nanotechnology is the ability to store data at the molecular level. Volumes of books can already be stored on a disk, and this improvement will continue with molecular storage of data.

DEVELOPMENTS IN NANOCOMPUTING

There have been countless approaches for handling the advent of molecular computing, and Computer scientists today have come up with several proposals to build components of nanocomputer. Some of these approaches will be summarized below, making note of the advantages and disadvantages in their implementation.

DNA Carbon Nanotubes and Nanowires

We know that a computer's building blocks can be a key

component to system efficiency and speed. An organic carbon based nanowire structure has been proposed by a group of researchers from Duke, NC State, and The University of North Carolina.

These researchers have focused on using the building block of the human body as the basis of building a computer. In this proposal, the top-down assembly approach of lithography would be replaced by DNA guided self-assembly. The DNA assembly uses strands of DNA to create tiles which in turn create a lattice structure. They focus on a waffle-like lattice of 16x16 nanometer size, and they attach carbon nanotube devices to the DNA tile scaffolding through a process they call functionalization, which is the process of adding a molecule to another material through a chemical reaction.

This circuit architecture does have a few shortcomings inherent in it. The self-assembly of the DNA lattices proposed does not produce a regular pattern which would insure a low level of defects. They have found that by minimizing the number of tags and producing repetitive structures is desired. However, by minimizing the number of tags, there is a limit to the amount of complexity a circuit can then have. The DNA self assemblies will not be able to have as much complexity as a CMOS transistor as a result.

In a shift away from the CMOS circuit, the DNA self assemblies are designed to have some level of defect tolerance. This is mostly due to the fact that the error rate in photolithography is quite low. However, self-assemblies will inherently have a higher number of defects in production. This particular proposal does not outline a method for handling defects, and that is a shortcoming of this presentation.

The proposed architecture for DNA self assemblies must handle limited node connectivity and random node placements. These limitations affect the performance in large scale control of the nodes. An accumulator based instruction set architecture is proposed to deal with the control and communications limitations. The accumulator-based ISA reduces the need for widespread coordination and communication among many components, since the only data dependence involves the accumulator, and instructions are processed in order. Each instruction within the accumulator makes up an execution packet containing operands, operations, and the accumulator value to form a sequence.

This approach supports parallelism through the independence of data, and by initiating several execution packets at once.

This proposal also puts forth a networking architecture for the execution packets. Three logical networks are proposed to handle three different types of information: memory requests, memory responses, and execution packets. Memory and execution are separated onto two different physical networks to avoid problems with deadlocking.

This research has mapped out some of the major hurdles they must clear and are thinking ahead to putting their fabrication together to make a system, but they are still in the beginning phases. Another paper produced by this group dealing with current research topics outline their experimental progress with carbon based nanotubes and nanowires. A 4x4 DNA tile has been produced for a variety of application computations, and they have demonstrated highly parallel computation via a set of DNA crossover tiles. They hope to fashion a simple circuit with nanotube control in the near future.

Chemically Assembled Electronic Nanotechnology (CAEN)

This approach is presented by Seth Copen Goldstein at Carnegie Mellon University. Goldstein believes that CAEN devices will be the cornerstone of the nanocomputer and replace the dominant silicon transistor. The devices he proposes are small, dense, low power and reconfigurable.

The proposed fabrication of these devices makes switches and wires using chemical self-assembly as opposed to the traditional lithography method which is currently used for CMOS chips. As stated by Goldstein, "The active devices are created where two appropriately functionalized wires intersect. By choosing the molecules that coat the wires appropriately, the intersection points can be ohmic contacts, diodes, configurable switches, etc. The important point to note is that no precise alignment is required to create the devices."

After proposing how these devices would be manufactured, Goldstein goes on to explain how a computer using these devices might be architected with his current fabrication plan. Cost and fault tolerance is the most important aspects of CAEN devices, and Goldstein explains that the proposed manufacturing process will result in the following architectural characteristics:

Transistors, or other three terminal devices, will probably not be part of CAEN devices.

Any deterministic aperiodic circuit will have to be created after the device is fabricated through reconfiguration.

Chips will have to work in the presence of a significant number of defects

Unfortunately these CAEN devices cannot create three terminal devices because precise alignment is required and this is not economically viable. CAEN devices would have to be limited to two terminals, and any processes which required three terminals (such as an inverter) would have to be built in conjunction with a CMOS device.

Goldstein believes that CAEN based devices can replace the silicon chip. These devices have the possibility of being reconfigurable, testable, and can be fault tolerant. A device called a negative differential resistance (NDR) can replace the traditional transistor circuit. Unfortunately, current research requires a great deal of voltage in order for this device to work. Further study will be required in order to produce a low power and low static device.

Despite its current limitations, the NDR is a key piece in Goldstein's proposed resonant tunneling diodes which are the building block of the molecular latch which will in his mind ultimately replace the transistor as we know it.

NANOBLOCKS AND NANOFABRICS

Goldstein has built upon the concept of Chemically Assembled Electronic Nanotechnology (CAEN) by outlining a reconfigurable architecture called a Nanofabric. This Nanofabric architecture could be used for factory-programmable devices configured by the manufacturer to emulate a processor or other computing device, or for reconfigurable computing devices.

A Nanofabric is a two dimensional mesh of interconnected NanoBlocks. A NanoBlock is a logic block that can be programmed to implement three-bit input to three-bit output Boolean function and its complement, as well as for switches to route signals.

This paper outlines how to build an AND gate, an OR gate, and XOR gate, and an ADDER. These "gates" as we traditionally are used to calling them, exist in what they call the molecular logic array (MLA) where all of the functionality of the block is located. The upside to the method outlined in the paper is that it is similar to commercial Field Programmable Gate Arrays (FPGAs), allowing these nano-gates to leverage current FPGA tools.

The MLA would be created using "directed assembly", which means that the atoms are arranged one by one. While this would allow computer makers to place components in a precise manner, the signals, which are routed using diode-resistor logic, would be degraded every time it goes through a configurable switch.

While a plan for defect and fault tolerance is presented, the inherent problem is with the nanowires lacking resistance. Goldstein contends that while this is a problem, it is still easier to implement defect tolerance than with CMOS because wires can be swapped and rows can be moved anywhere within the NanoBlock without affecting the circuit. CMOS currently has little defect tolerance, and chips with a high error rate are discarded. The cost of directed assembly may be prohibitive as well and not economically viable.

Testing the Nanofabric proved to be difficult because it is not possible for the components to be tested in isolation. After production of the fabric, a defect map is created to make the fabric into a viable circuit. The fabric is self-testing, and will provide fault free regions of the fabric.

Other researchers at the University of Massachusetts are looking to use Nanofabric to create integrated circuits which can be tailored towards the application domain. They are named NASICs, which stands for Nanoscale Application-Specific Integrated Circuits. This research is built upon the Nanofabric proposal by Goldstein. In an effort to build these NASIC tiles, the researchers at UMass have presented a solution to a difficult problem. As they state, "A key issue when building NASIC tiles is the low effective density of latch circuits. This is due to the difficulty of building sequential circuits on a nano-grid.".

In order to get around this limitation, they have developed a nano-latch, which allows for temporary storage along a nano-wire in a circuit. This research gives promise for improving nano-device density, scalability and reliability

Quantum Cellular Automata

A Quantum Cellular Automata (QCA) is another plausible nanodevice. This is a design which is based upon a theory of using electrons and their energy states for computing. This is a topic which has been studied for over 40 years, back to the work of von Neumann. Rather than being an individual device of a computer circuit scaled to a nanometer size, the QCA is a complete entity. As described by researchers at Notre Dame, "quantum-dot cellular automata (QCA), parallels conventional computing in using a binary representation of information (not qubits) yet realizes functionality in ways which are well-suited to the properties of single molecules - using molecules not as current switches but as structured charge containers."

The QCA is a single molecule which will contain a representation of the binary values 0 or 1 based upon the static electric charge of the electron. The cells have not been found at this point in time to be viable at room temperature, but rather only at very cold degrees in the Kelvin range. Needless to say, operating at such low temperatures is not practical for any working device, and this is a serious limitation to the usefulness of QCAs.

The QCA fabrication does differentiate from other devices

that we have seen because the fabrication uses lithography. "Molecular QCA cells must be attached to surfaces in ordered arrays. We are using electron-beam lithography to burn narrow "tracks" along surfaces."

The temperature limitations of Quantum Cellular Automatas have been given a new perspective through a study of Magnetic QCA described by Cowburn and Welland. Their presentation uses interacting submicron magnetic dots to perform logic operations and propagate information. Although these devices will not be in the nanometer range, they will use less power than current CMOS.

CMOL

Some researchers are presenting a possible bridge from CMOS to the next generation of molecular computing by formulating a hybrid circuit. One such hybrid circuit is dubbed a CMOL, a play on the combination of CMOS and molecular. The CMOL is "a circuit combines an advanced CMOS subsystem with two, mutually perpendicular, arrays of parallel nanowires and similar molecular devices formed at each crosspoint of the nanowires."

This approach is important in several ways. First, it's unlikely that the computing industry will not be able to make the leap from its top-down lithography methodology to a bottom up approach using nanowires and integrating with some molecular components without some intermediate step.

The CMOL maintains the stability of the silicon chip and takes a small step forward by using bottom up fabrication. The CMOS circuit maintains the level of functionality that we are used to. As mentioned in the previous chapter, the system bus can be an area in which slowness can occur. Experiments with the CMOL circuits have operated faster than regular computers with better power dissipation.

The creator of the CMOL hybrid circuit claims, 'The development of CMOL technology (especially the high-yield molecular self-assembly) will certainly require a major industrial effort and substantial time period, probably not less than 10 to 15 years. However, this timing may be still acceptable to prevent the impending crisis of Moore's law, provided that we start right now."

As if the urgency wasn't any more obvious.

Markov Random Network

Researchers at Brown University have a proposed architecture based on Markov Random Fields (MRF). A MRF is a concept based on Markov chain, which is a "graph of stochastic variables, where each variable has the property that is independent of all the others (future and past) given its two neighbors" A Markov Random Field network allows circuits to behave in a relatively independent fashion, allowing for re-configurability and a good level of fault tolerance.

The researchers feel that carbon nanotubes are one of the most promising devices which have been built at the Nanoscale.

Some issues are inherent with these nanodevices though; these researchers point out that the use of carbon nanotubes circuits will increase failure rates and bring about heat dissipation issues. In order to deal with possible failure rates upwards of 10% and heating issues at the thermal limit, it is necessary to configure a network of circuits which can handle a high fault level and can somehow dissipate heat throughout the circuit network.

It has been found through experiments that the use of a Markov Random Field Network in conjunction with the Gibbs formulation for dissipation of heat can be a viable solution, because as the authors state, "its operation does not depend o perfect devices or perfect connections. Successful operation only requires that the energy of correct states is lower than the energy of errors"

Cell Matrix

The Cell Matrix architecture is presented by the Cell Matrix Corporation in Salt Lake City. This piece of research differs from the previous sections in the sense that this is a proposal of a true architecture as opposed to a design of a nano-sized component.

This architecture is currently being designed and tested in the silicon chip domain, but all research, configuration and design is for the coming of nanostructures. Currently, the Cell Matrix Corporation is developing an atomic cell unit, which is repeated to form a multidimensional matrix of cells to make a highly scalable architecture, which differs from other proposals discussed thus far. The architecture is similar to the field programmable gate arrays (FPGAs) like the Nanofabric, and it also has similarities to the cellular automaton proposed by Von Neumann.

Several aspects of a cell matrix distinguish it from a Von Neumann cellular automaton. First, the cell matrix has the same three dimensional nature. The cell matrix can only be in a certain number of states it can be in at any given time, just like a cellular automaton. Second, a cell matrix is fine grained and reconfigurable, and third it is programmed like a digital circuit, and not like a cellular automaton.

The benefits which cell matrix architecture provides are many. It can efficiently handle a large number of switches due to the fact that its controlling system is highly scalable, and can handle any addition of cells to the system. This architecture also promises to provide highly parallel large scale computing. This is the result of the fact that each cell contains all necessary functionality, and can compute across the matrix at any point in a parallel fashion. This group has demonstrated this highly parallel computing ability through an experiment with a search space problem.

The cell matrix architecture relies on a very simple hardware definition mixed with a complex programming of each individual cell. Simple logic functions can be handled by individual cells, whereas more complex functions are handled by a collection of cells, which are spatially close to each other. In these collections, cells are set up to perform a subset of the work of the entire circuit. These cells communicate with one another in an asynchronous fashion. Each cell also has the ability to reconfigure itself, allowing for self-testing and fault tolerance, although experimentation and testing in this arena is still taking place.

The Cell Matrix Corporation believes that their design can provide dynamic circuitry, allowing each of the cells to change their original behaviour or the behaviour of cells around them, or to migrate into new areas within the matrix. They also see this as being a mechanism to handle faulty cells in order to make their circuits fault tolerant.

Defect tolerance is always an important issue with any computer architecture due to the cost which defects add to the final cost of a circuit. Due to the fact that silicon chips are discarded if their fail rate is too high, a new architecture can find cost savings in total manufacturing costs if the new circuitry can fin a way to be fault tolerant by working around bad cells. The Cell Matrix has implemented defect tolerance and self testing in their architecture. They have designed a test driver which has the following goals:

Permit reporting of faults with a high resolution

Permit access to a region despite failed regions near it.

It should be easy to extend to a decentralized, parallel, distributed fault testing process with as small a footprint as possible.

Have a driver which can share the hardware with other tasks so it can perform its functions on subcomponents of a critical system while that system is running.

The Cell Matrix architecture outline is well planned and several experiments on digital circuits have been promising. However, no experiments have yet been performed on any microscopic components. In theory the layout may be scalable and successful; more incorporation of current nano-component research should be considered.

NanoPrism

It has already been determined that fault tolerance, while something not so prevalent in the construction of silicon chips, will need to be an important part of any architecture presented for a nanocomputer. One way in which many manufacturers provide better reliability is by implementing techniques to increase redundancy. However, it is known that adding redundancy cannot always increase reliability of a device due to the fact that the redundant device can also contain faults. A paper from Virginia Tech is attempting to determine the level at which fault tolerance and redundancy can produce a reliable nano-architecture.

As stated by the researchers, "The questions we try to answer in this paper is, what level of granularity and what redundancy levels result in optimal reliability for specific architectures. In this paper, we extend previous work on evaluating reliability-redundancy trade-offs for NAND multiplexing to granularity vs. redundancy vs. reliability trade-offs for other redundancy mechanisms, and present our automation mechanism using the probabilistic model checking tool PRISM.".

The NANOPRISM tool is based on another probabilistic model checking tool called PRISM built at the University of Binghamton, which is designed for conventional CMOS architectures.

The NANOPRISM tool will be a valuable resource for building a new architecture for a nanocomputer. Redundancy can be a very important yet delicate technique to employ, due to the fact that employing too much can cause a lessening of reliability, not to mention the fact that there are several levels of granularity at which redundancy can be implemented.

Because this tool automates the checking of a defect tolerant architecture, areas in which trade off levels are achieved can be identified much more quickly. The number of circuits and cells are going to increase exponentially as nanodevices scale up to the level they are supposed to be at. If we currently have almost 1 billion transistors on a single silicon chip, the number of devices we can just imagine the numbers present in a nanocomputer.

A PROPOSED NANOCOMPUTER ARCHITECTURE DESIGN

After looking at all of these different nanodevices and proposed networks and architectures, how can we determine what is the best direction to go in order to create the optimal computer architecture for the next generation? The first thing to look at is what will be the major issues and limitations with the devices we wish to use as the basic building blocks for our computer. The focus will not necessarily be on how these devices will be built whether that is through lithography or self assembly, it is immaterial, and concern about cost will be factored in, but not necessarily a focal point. Rather, we are more concerned with what qualities the finished product and the surrounding architecture will need to possess in order to be successful. What are some qualities are interested in?

Fault Tolerance,
Self testing/ Re-configuration,
Heat Tolerance and Power Dissipation,
Independent and parallel processing,
Improved communication, handling current pervasive bottlenecks in interconnections,
Scalability.

We will use these important qualities as a guide to what each nanocomputer component will contain, as we lay out our proposed nanocomputer architecture from the bottom up.

Chip Design

The chip design of the nanocomputer would be best suited to have a carbon base which is both self-testing and reconfigurable. Because we know that faults will be at levels upwards of 10% for any chip built in the nanometer range, it will be necessary for these features to be part of any solution we choose. Also, it's important to remember the breadth of our failure rate.

A Nanocomputer will contain not just one billion but several billion chips. If we have a computer with 5 billion chips, we will have 500 million chips which are defective with a 10% error rate. It will be an important cost factor to be able to repair these chips after they have been fabricated. Of the research we have gone over thus far, the best proposed replacement for the CMOS chips is the NanoBlock, based on the FPGA, the Field Programmable Gate Array. These blocks are carbon based. These devices have been created from a bottom up assembly; they are fault tolerant and reconfigurable. Bad devices or blocks of devices can be swapped out if necessary. Another advantage to the NanoBlock is that it is chemically assembled, allowing more flexibility with reconfiguration. The devices can be manipulated so that changes can be made to improve any faults which may make an entire block unusable.

The argument for NanoBlocks also stems from the need to insure heat and power dissipation. The NanoBlocks are low power devices, and this makes it a good candidate as a base for building a nanocomputer.

Instruction Set Design

Very little research in nanotechnology has addressed the future of instruction set design. We know that our hardware systems will become more complex and more powerful. We can infer that the instruction set design should then move back from a reduced instruction set (RISC) design to a more powerful complex instruction set (CISC). This way, we can take advantage of the improvements in the hardware and write less complex compilers. The instructions can also perform more work in a single clock cycle, improving the speed operations are performed in the system. However, the one architecture that we have looked at which took instruction set design into consideration only used a simple accumulator.

One would think that a more complex design would be beneficial and more productive, but it is possible that a simple approach will be the best way to manage so many million components. RISC was chosen as the instruction set for this generation of components due to cost, and it's conceivable that the next generation will choose a similarly simple instruction set due to size constraints and sheer numbers. The instruction set will presumably be one of the last design components to consider, but designers must take it into consideration when building components.

Interconnections and Communication Points

Interconnections are another vital piece to consider. If there can be devices in the multi-billion in a nanocomputer, these nano-connectors will also number in the multi-billion and will need to be efficient and allow for both local and global communication. The development of nanowires as communication pathways between neighboring cells is the most well developed form of nano-sized interconnects. The carbon based nanowires are the most promising format because of their stability.

These devices also allow for self-testing, reconfiguration, and self-assembly. Each of these qualities makes it an attractive component in terms of costs, reliability, and practicality. It is also presumed that the reduction in size of these wires can improve density and we can move more data through interconnections at a faster rate. This can reduce starvation which can sometimes occur in a system due to bus latency and lack of bandwidth. While carbon nanowires have been produced in the lab, the nanowires which we must use in our nanocomputer must improve their resistance. Arranging these nanowires in a Markov Random Circuit as one group of researchers has proposed would be a valuable arrangement, and would be an important algorithm to employ in a nanocomputer.

Main Memory, Multiprocessing and Instruction Level Parallelism

Memory management is and will continue to be a very difficult topic facing all computer designers. Virtual memory demands which are currently limited by space will be unlimited in a nanocomputer architecture. Memory will be more plentiful, cheaper and faster in a nanocomputer, and we can expect to move from the 1 gigabyte of memory standard on modern computers to numbers two to five times more powerful. However, these advantages will bring a new host of issues. Instruction level parallelism may be more difficult because there will be so much more data to handle. Algorithms will need to be developed to insure that critical sections of code are executed properly. However, we do know that we will have many more CPUs which we can cluster together with much less space and cost requirements.

It is likely that instruction level parallelism will fall away to a new paradigm of handling large scale multiprocessing.

Perhaps machines will be split apart so that certain CPUs are specifically for certain jobs, or certain tasks within a computer. Nanocomputer research has not reached the point of coming up with a memory management scheme, but the chances that it is exactly as we handle memory today is unlikely.

Storage

A nanocomputer will have the possibility of an unlimited amount of storage space. Currently we have computers which can hold up to 150 gigabytes of information. We expect a nanocomputer to be able to hold terabytes of information. These storage disks will also be easily searchable and can also be partitioned. Presumably we can hold volumes of data, and backup tapes or requiring repositories to be placed on a network of computers to make up a server can be a thing of the past. Smaller embedded devices will be able to hold much more data as well. Our entire medical history can exist on a key, or entire music collection can exist on a device the size of a credit card or smaller.

Conclusion

The future of nanotechnology will bring exciting change to the computing industry. We are promised machines which are faster, cheaper, and smaller. This promise will come with incredible challenges and sacrifice, and the computing industry must be flexible and agile enough to meet the demands of manufacturing these new devices and designing the machines which will contain them. The industry must also change the current manufacturing processes and current design paradigms to meet the differences between this new generation of devices and our generation. These new devices will require a completely new approach in manufacturing, moving from lithography top down design to bottom up design.

It may also require self-assembly of components in order to be economically viable. This will require a new set of manufacturing tools and processes. Computer manufacturers will have to change their plants and assembly lines completely in order to build these new components. A nanocomputer will also require a new instruction set, and innovative ways to handle the new challenges when dealing with molecular components must be considered. Hardware will be less reliable and self-testing and self-configuration must be built in to any nanocomputer. This may make manufacturing more expensive at first, and more time consuming.

The end of hardware improvements predicted by Gordon Moore is upon the computing industry, and lab experiments to build the next generation of hardware must take shape to keep pace with the demanding computing needs of the world. Valuable research has been made in many areas, but we are still in the primitive phases. The industry has about ten or twenty years left before this paradigm shift must occur. It can be gleaned that we could experience some setbacks in the new generation of devices were they may not perform as reliably or as quickly as we are used to with CMOS based computing devices. However, one could say that it would be worth taking a few steps back in order to move many hundreds of steps forward, which is what nanotechnology is promising us.

Figures and Tables

Below are some pictorial representations of some of the nanodevices and architectures we have discussed in this paper.

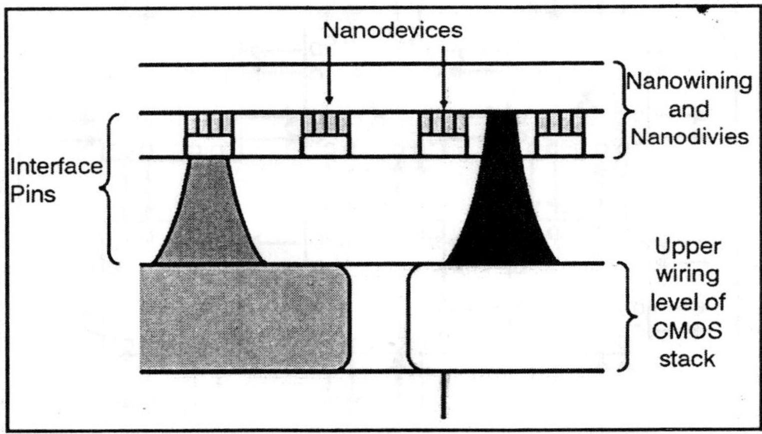

Fig. Structure of a CMOL Circuit

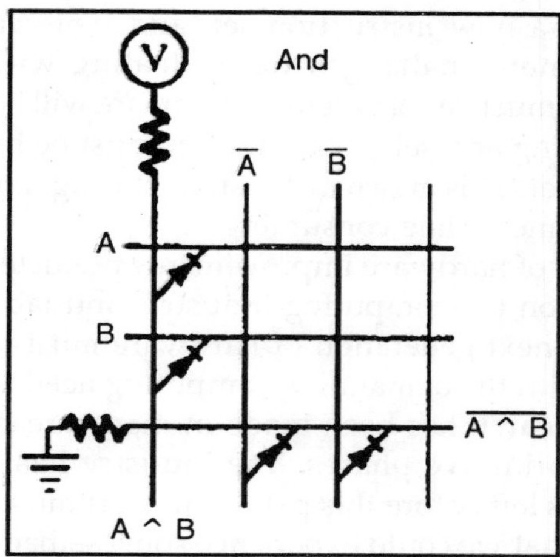

Fig. A two input AND gate Implemented in a CAEN Grid

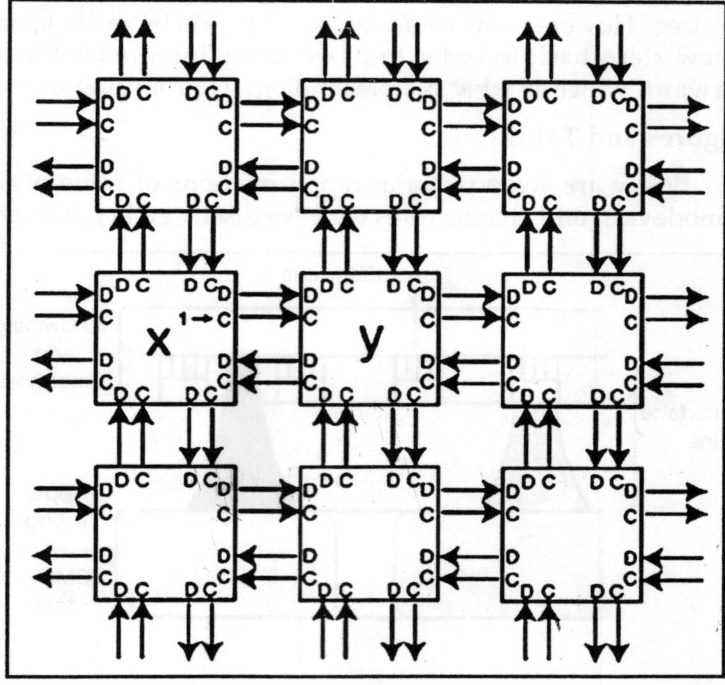

Fig. 3x3 Cell Matrix

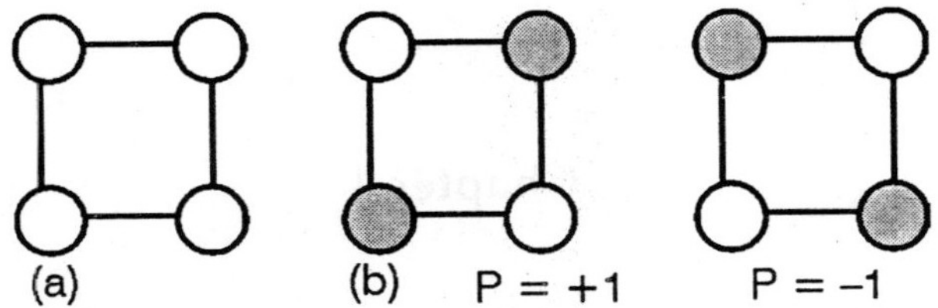

Fig. QCA Four Dot Cell

Chapter 4

Rod Logic and Thermal Noise in the Mechanical Nanocomputer

Molecular electronics embraces both molecular electronic materials, in which devices consist of large aggregates of molecules, and molecular circuitry, in which distinct devices are themselves of molecular scale. Molecular circuitry is closely related to molecular machinery in scale, energy requirements, and required technology base. Both circuitry and machinery can support molecular computation; the focus here will be on the latter.

Molecular machinery and circuitry (save in their simpler instances) will be forms of *nanotech*nology, technology based on a general ability to build systems to complex, atomic specifications. Nanotechnology will be based on *assemblers, molecular* machines that guide chemical bonding operations by manipulating reactive molecules.

The concept of molecular machinery was latent in the earliest efforts to describe molecules as mechanical systems. The molecular theory of gases marked an early milestone; the recognition that biochemistry and molecular biology rest on molecular machines marked another. Suggestions for building molecular machines, however, have been sparse, and the concept of using molecular assemblers to direct synthetic reactions is apparently quite recent.

Richard Feynman's 1959 talk on miniaturization foreshadowed the concept of the assembler, though it lacked any clear discussion of molecular machinery, of directing chemical bonding, or of reliability in the face of thermal noise.

Feynman observed that 'The principles of physics, as far as I can see, do not speak against the possibility of maneuvering things atom by atom," and went on to remark: "But it is interesting that it would be, in principle, possible (I think) for a physicist to synthesize any chemical substance that the chemist writes down. Give the orders and the physicist synthesizes it. How? Put the atoms down where the chemist says, and so you make the substance."

The case for assemblers can be made most compactly through comparisons to existing biochemical machinery. Enzymes, ribosomes, and the genetic system demonstrate that molecular machines can exist, that they can position and manipulate reactive molecules, and that they can make and break specific chemical bonds under programmed control. These biochemical machines demonstrate lower bounds to the capabilities of more general molecular machines.

These demonstrated capabilities indicate that protein engineering will offer a way to construct novel molecular machines, including devices resembling simple industrial robot arms-having perhaps two degrees of freedom-and able to position reactive molecules with respect to a molecular workpiece. These could serve as primitive assemblers.

The chief problems of synthetic chemistry are controlling reaction sites, providing activation energy, and providing free energy to drive otherwise unfavorable reactions. The use of even primitive assemblers to position reactive molecules will improve control of reaction sites, permitting the construction of molecular structures of a complexity well beyond reach of modern synthetic chemistry.

Among these structures will be components for better molecular machines. Thus, however the first assembler may be built, we can expect to see an iterative process in which assemblers are used to build better assemblers. The use of strong, rigid materials in assemblers (as opposed to flexible polymers) will aid the use of mechanical energy to drive chemical reactions; other components will do likewise for the use of electrical energy. Together, these advances will make possible the synthesis of an extremely broad class of structures, which may be expected to include virtually all molecular structures of engineering interest. Protein engineering is already being pursued . Other paths to assemblers seem possible: these include further development of non-protein supramolecular chemistry and the development of micromanipulators based on the positioning and sensing technologies demonstrated in scanning tunneling and atomic force microscopy.

The nature of these steps is of only short-term importance: All paths lead ultimately to molecular machines able to build systems to complex, atomic specifications, including systems built from separately assembled parts. It is thus of interest to consider what sorts of computers can be built with this fabrication technology.

NANOCOMPUTERS

Nanocomputers based on suitable electronic effects should be far faster than those based on machinery: logic operation times may be in the femtosecond range for the former, but will be in the multi-picosecond range for the latter. Mechanical nanocomputers, however, can be modeled using newtonian mechanics and empirical force fields , rendering analysis comparatively straightforward.

If one aims not to design the best possible computer, but to set firm lower bounds on future capabilities, mechanical nanocomputers are attractive targets for study.

Moreover, the results presented here suggest that they can operate near the apparent limits of volume and energy consumption for computers having structures analogous to those in current use.

Work on this concept has produced schemes for signal transmission, logic gates, finite-state machines based on programmable logic arrays, data storage registers, power supplies, power transmission systems, and means for input and output of data to conventional microelectronics.

This paper focuses on the rod logic used to implement signal transmission and logic gates, describing its structure and dynamics with special attention to the problems posed by thermal noise.

The Rod Logic Approach

In rod logic, flexible tensioned rods serve as signal transmission elements (like conductors in integrated circuits, but requiring a significant turning radius at bends). Rod displacements serve as signals (like voltages in wires). Knobs in these rods can block the displacement of other rods; these knob assemblies, or *locks,* serve as logic elements (like field-effect transistors). Knobs are of two types: gate *knobs,* which can block, and *probe knobs,* which can be blocked.

A lock implements a conditional restraint on motion: the rod attached to the probe knob can be pulled in the direction indicated by the arrow if and only if the obstructing gate knob has been pulled from its path. If a gate knob is ordinarily out of the way, but can be pulled into a blocking position, the result is a logical not operation. (With this as a normal mode of operation, allowing free use of both default-open and default-closed gates, then the resulting rod logic will resemble CMOS logic more closely than it does NMOS.) Figure shows a rod with three input locks and three output locks in an abstract representation.

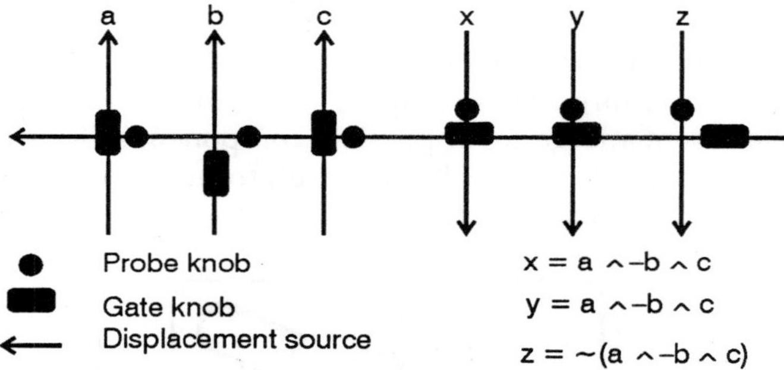

Fig. Schematic of a three-input, three-output rod, with Logical Description

Rods slide through channels in a close-fitting matrix, forcing essentially longitudinal motion and thus allowing gate knobs to block probe knobs by blocking their channels. The relationship of knobs, rods, and the matrix is shown schematically in Figures.

In computation, the sequence of operations is as follows: First, the rod containing the gate knob rod is pulled; if it is in an unblocked (or "1") state, it moves; if it is in a blocked (or "0") state, it cannot.

Second, the rod containing the probe knob is pulled; if it is not blocked elsewhere, then it will move if the gate knob is in a non-blocking position. Third, after a delay (during which gate knobs on the rod containing the probe knob can affect operations elsewhere) the pulling of the probe knob is reversed; if it had moved, it is reset. Fourth, the pulling of the gate knob is likewise reversed, restoring the initial conditions.

Key issues in rod logic include the design of the rods, knobs, and matrix, the nature of logic states and transitions between them, and the effects of friction and of thermal noise. These issues are interrelated.

In particular, the design of the rods, knobs, and matrix must meet constraints stemming from friction and thermal noise. At one extreme, parts could contain many atoms and move slowly, resembling familiar macroscopic machinery; this possibility shows that rod logic can be made to work on some scale.

The design choices presented here are closer to the extreme of compactness and speed, but are nonetheless selected from a large set of possible component sizes, shapes, and arrangements. These choices result from several iterations of analysis and redesign, trading off performance and ease of design and analysis.

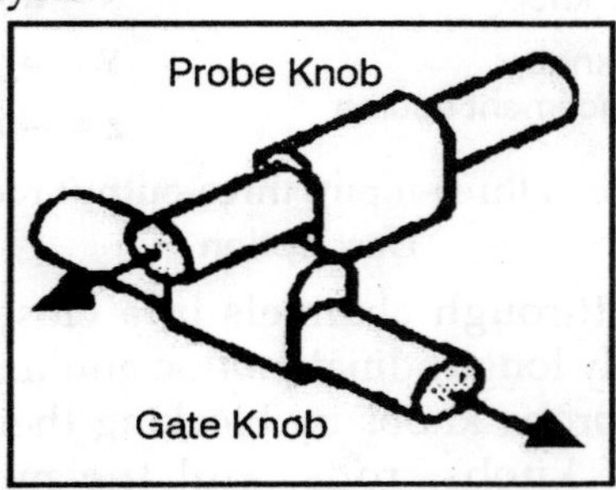

Fig. Schematic of Lock Components

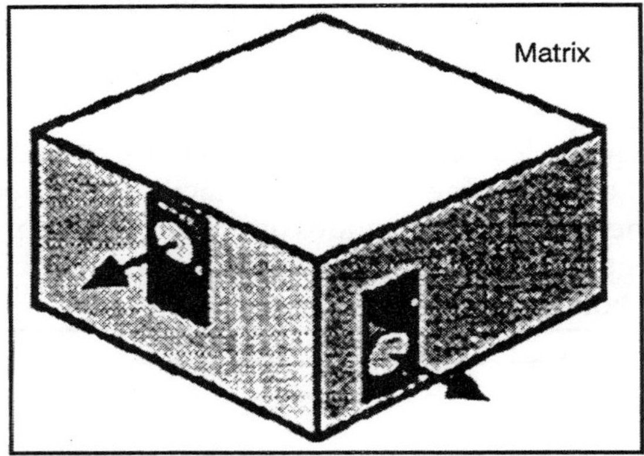

Fig. Lock Components in Channels

ROD LOGIC ELEMENTS

Inter-knob Segments

To minimize bulk and maximize speed, one seeks designs based on light, stiff parts of minimal size. These considerations favour use of light, covalently-bonded elements in moving parts that are one atom wide.

For the inter-knob sections of rods, one choice seems outstanding: segments of carbyne, linear chains of carbon atoms linked by alternating triple and single bonds. Since it is one atom wide, carbyne is at or near the limit of slimness; its diameter is about 0.30 nm (this radius, like others in this paper, is based on atomic separations yielding atom-atom contact forces of about 0.1 nN).

Single-bonded carbon (i.e., diamond) is renowned for its strength, but the single bonds of carbyne are almost as short and strong as double bonds (0.1373 nm vs. 0.1337 for double bonds and 0.1541 for single; all bond parameters in this paper are from or unless otherwise noted). The strength of a rod will be well in excess of the 6 nN strength of a C-C single bond. Further, the linear structure of carbyne allows bond bending to make no contribution to elongation under tension. Accordingly, carbyne is strong and stiff. (Here, stiffness means elastic modulus in tension; rods are quite flexible in bending or in compression.)

Knobs

Choices are less clear where the knobs are concerned. Benzene-like pyridazine rings are attractive for knob bases; like benzene, pyridazine provides opposed bonding points for carbyne segments, but the pair of nitrogen atoms reduces the size of the bottom of the knob.

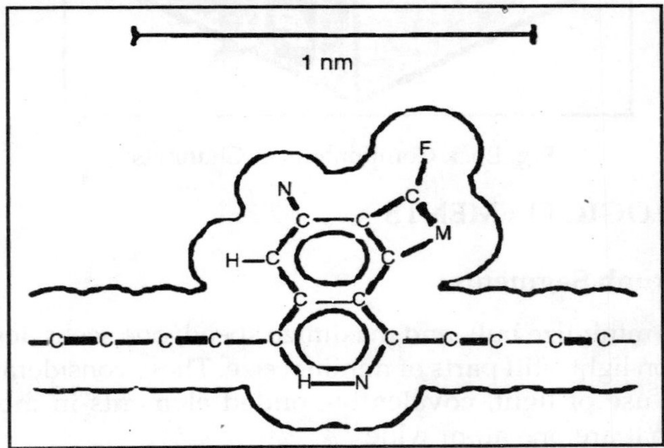

Fig. Knob Structure (Note Carbyne Segments)

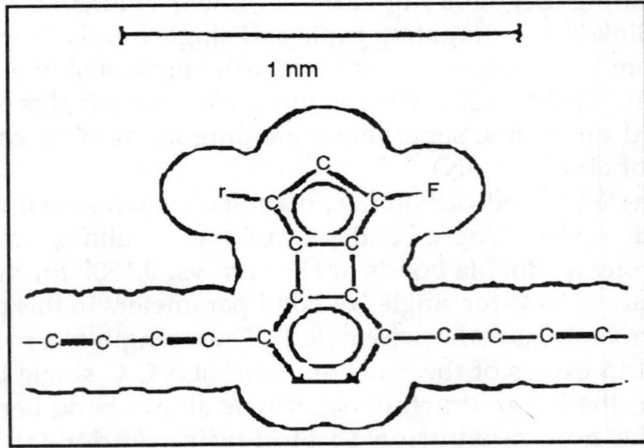

Fig. Ate Knob Structure note Carbyne Segments)

Figure shows the assumed probe knob design: a fluorine atom forms the chief mechanical contact to a blocking gate knob, and the rest of the structure was chosen to hold this atom in a forward position, in plane, at a suitable height, and with adequate rigidity. Figure shows the assumed gate knob design: this structure, incorporating a fluorinated thiophene ring, was chosen for its height, length, planarity, and comparatively smooth width profile near the F-F axis.

In both of these designs, the height of the structure serves to place the upper portion of the knob well above the track.

The probe knob holds its fluorine atom 0.57 nm from the carbyne axis; its total height is 0.97 nm. The gate knob holds its fluorine atoms 0.43 nm from the carbyne axis; its total height is also 0.97 nm. The gate knob's length is important in thermal noise calculations: the F-F internuclear distance is 0.54 nm, and the length (measured from 0.1 nN van der Waals surfaces) is 0.80 nm. (All van der Waals parameters in this paper are derived from and.)

Rod Parameters

The stretching modulus of the rod is also important in thermal noise calculations a high value is preferred.

Rod modulus depends on the spring constants of rod components.

Modeling the pyridazine base of a knob as a benzene ring, and calculating from force constants determined by Pulay *et al.*, the knob-base spring constant is about 220 N/m. Rogers *et al.* give the spring constant of a C-C triple bond as 1560 N/m. Using this value and a ratio derived from the diagonal force constants in Popov and Lubuzh, the spring constant of a C-C single bond between two triple bonded pairs is about 824 N/m. Correlations between C-c bond lengths and stiffnesses suggest that the spring constant of the bond between the pyridazine ring and a triple-bonded pair is about 680 N/m.

To reduce lateral vibrations, rods will operate under tension, here chosen to be 2.5 nN. As tension increases bond lengths, it reduces bond stiffnesses owing to nonlinearities in the interaction potential. These effects have been estimated by fitting a Morse potential to zero-tension bond length and stiffness data, and then evaluating bond lengths and spring constants at the given tension.

The compliance of a benzene ring is dominated by bond bending and bending interactions become stiffer with increasing displacement ; knobs have been modeled as linear springs and the expected increase in stiffness with tension neglected. Transverse vibrations of the rod reduce the effective modulus. If all the above spring constants were infinite, rods at 300 K and 2.5 nN tension would have a finite modulus of about 1500 nN from vibrations alone. A major reason for tensioning rods is to reduce this effect. The average rod modulus will be affected by the number of triple-bonded pairs per inter-knob segment; it seems that four pairs will suffice. With this rod configuration and the parameters and corrections just described, the rod modulus is about 89 nN.

These calculations also yield the length of a tensioned subunit consisting of one knob and inter-knob segment, 1.50 nm. The associated mass per unit length is 3.0×10^{-16} kg/m for a series of probe knobs and 3.2×10^{-16} kg/m for a series of gate knobs. Taking the mean of these mass densities together with the given modulus, the speed of (longitudinal) sound in a rod is

$$[(89 \times 10^{-9}\ \text{N})/(3.1 \times 10^{-11}\ \text{kg/m})]^{1/2} \text{ » } 17{,}000\ \text{m/s}.$$

Locks

Figures illustrate the geometry of gate and probe knobs in their intersecting channels. Each intersection forms a lock. Note that whether a lock is open or closed depends both on the position of the gate knob, determined by the momentary logic state, and on the cant of the probe knob, determined by the type of the lock-that is, by the fixed structure of the matrix and track.

The volume per lock in a close-packed array is determined by the spacing of gate and probe knobs on their rods (1.50 nm) and on the total height of the lock assembly. Assuming that the fluorine atoms on the gate and probe knobs are aligned for a direct collision, and that one carbonatom diameter is allowed between the channels of stacked locks, the height of a lock is 1.81 nm. This yields a volume per lock of 4.1 cubic nanometers.

Other Elements

Figure shows a schematic illustration of a rod with the remaining necessary moving parts. These include:

- Two springs. One is a stiff drive spring and the other an *anchor spring* soft enough to apply an approximately constant tension of 2.5 nN to the gate segment over an extension of about 0.84 nm. (The value of 2.5 nN is chosen to satisfy thermal noise constraints;) These springs can be fairly bulky yet add little to the volume per lock in a long rod. Since it is easier to make molecular structures compliant than to make them rigid, and since many designs are possible, no spring design is specified here. Like knobs, springsegments can be covalently bonded to rods.

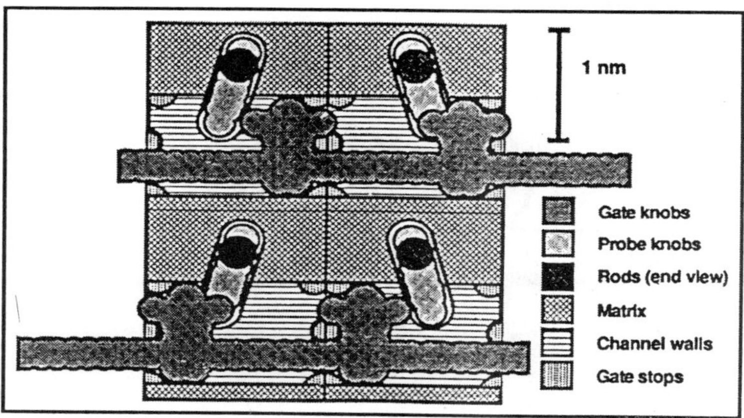

Fig. Four Locks viewed in section along two Channels Containing Gate Knobs.

Left to right top to bottom, the locks are in the states 1, 0, 0, 1.

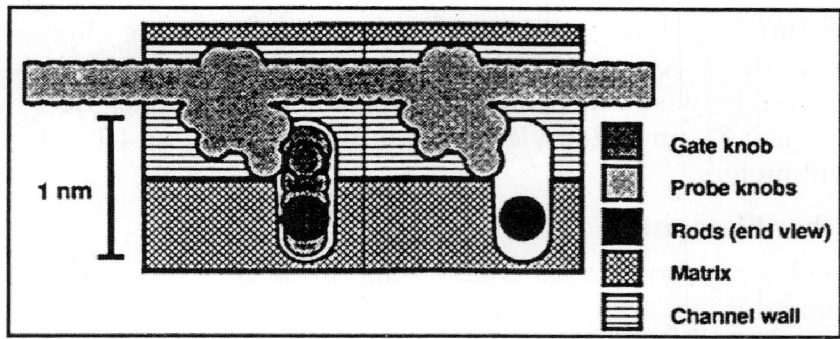

Fig. Two locks viewed in section along the Channel Containing their Probe Knobs.

Left to right, the locks are in the states 0, 1.

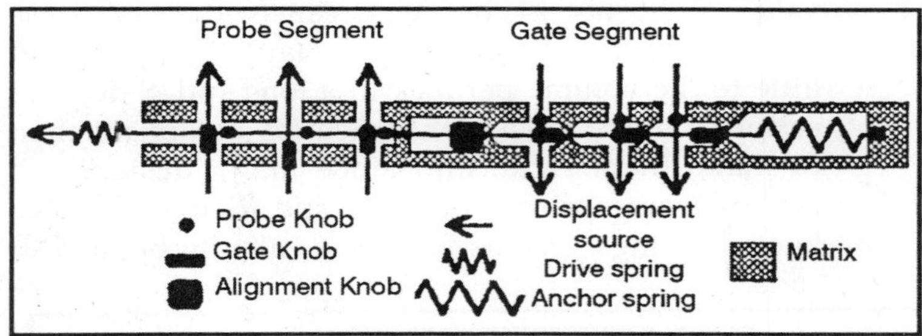

Fig. Schematic Diagram of Rod with Additional Moving Parts

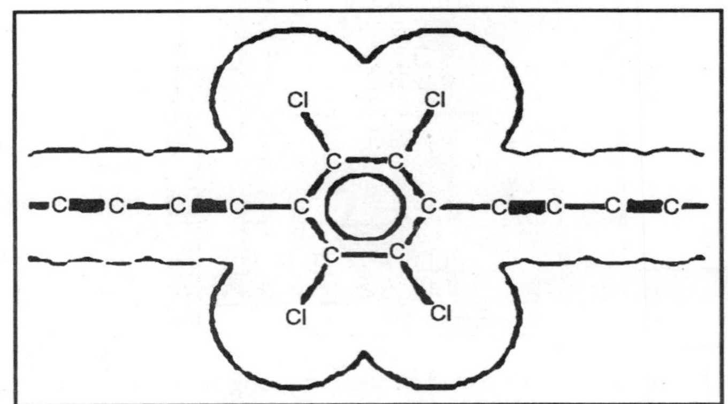

Fig. Alignment Knob Structure

- An *alignment knob*. A proposed alignment knob

structure is shown in Figure. This knob interacts with two barriers in its channel, the *alignment stops* (see .

- A *displacement source.* This provides properly timed displacements at the far end of the drive spring; the nature of this mechanism (which provides both power and timing for the rod logic system) is beyond the scope of this paper.

Chemical Stability

The stability of the inter-knob segments and of the strained, unsaturated ring in the probe knob demand attention. Here, one must distinguish between instability due to intermolecular reactions and instability due to intramolecular reactions.

Intermolecular reactions require intermolecular collisions, which here are highly constrained by the matrix. If the matrix surface is designed to be non-reactive, and if probe knobs cannot react with gate knobs (their constrained orientation prohibits an otherwise conceivable Diels-Alder reaction), then no intermolecular reactions can occur. Intramolecular reactions (such as bond breakage or rearrangement) would require either weak bonds or low potential barriers between strongly bonded states. Neither condition holds in the structures chosen, which should therefore be quite stable in this environment.

THE MATRIX

General Characteristics

The matrix includes all the non-moving parts of the rod logic system; its parts serve as a casing and as bearing surfaces. If the atomic number-density of the matrix is midway between those of polyethylene and diamond, and if the matrix occupies all the volume outside the rod channels, then the number of matrix atoms per lock is roughly 400 (vs. 44 atoms in the associated rod segments). With so many atoms, the number of possible arrangements is vast; no specific, comprehensive design is proposed here.

Given that the matrix consists chiefly of elements near carbon on the periodic table, arranged with a structurally sound, three-dimensional pattern of bonding, the matrix as a whole will be strong and stiff. Its three-dimensional nature will make its compliance negligible compared to those of the long, slim rods. Accordingly, the matrix is treated as a fixed frame of reference in the thermal noise model. The matrix is assumed to contain no extraneous atoms. Gas molecules can be excluded by sealing the outer surface of the rod logic system.

Tracks and Stops

Given a suitable channel geometry, the only matrix atoms that interact strongly with rod atoms will be those in the tracks and stops. The gate stops consist of atoms that protrude into gate-knob channels, limiting gate-knob motion to a certain range, as indicated by Figure.

Tracks consist of rows of atoms positioned as shown in cross section in Figure. A pair of rows lies on each side of the channel, with the members of the pair separated by about 0.3 nm and the pairs separated by about 0.45 nm. If the atoms in each row are 0.24 nm apart, then the number of track atoms per lock is about 50.

Track atoms have strong, repulsive van der Waals interactions with the atoms of the knob base rings and inter-knob segments. These interactions clamp the rod firmly, allowing only small lateral motions. A twisted track will couple longitudinal rod motion to rotation of knobs about the longitudinal axis; this is responsible for the varying cant of probe knobs shown in Figure.

Fig. Rod with (Gate) Knob and Track, viewed in Section Across Channel

Matrix Synthesis

The matrix is an essential part of a rod logic system. Its complexity and the general specifications just outlined put it clearly beyond reach of synthetic techniques that lack mechanisms for the general control of the placement of reactive molecules, and equally beyond reach of manufacture by machining bulk materials. Thus, the matrix (more than the moving parts) illustrates a useful structure whose synthesis demands assembler capabilities.

LOGIC STATES AND BASIC ROD DYNAMICS

Logic States

A lock is in a 1 state if its gate knob, at equilibrium, does not block its probe channel. Thermal noise fluctuations will (at worst) block the probe knob only momentarily, then allow it to pass. If the system's timing allows for such delays, these fluctuations will cause no errors.

A lock is in a 0 state if its gate knob, at equilibrium, blocks its probe channel. Thermal noise fluctuations cannot be allowed a significant chance of unblocking the channel even momentarily, since this would allow the probe knob to pass, causing an error in the logic system.

The lock configuration shown in Figure 6 reflects this asymmetry between 1 and 0 states. A small displacement can momentarily block the probe channel in the 1 state, but in the 0 state, the gate stops prevent small motions from unblocking the probe channel. Instead, the gate knob must move across the channel before it can cause an error state. The probability of this larger motion is analyzed in Section 7.

Logic systems may contain regular arrays of locks, as suggested by Figures and by analogies to integrated circuit designs. In such an array, a null lock may be constructed by omitting the tops of what would otherwise function as gate and probe knobs. In effect, these locks will always be in the 1 state.

A rod of the sort analyzed here is schematically illustrated in Figure. It consists of a series of probe knobs in a probe segment, each corresponding to one input lock. If all input locks are in the 1 state, the rod is in a I state, and is free to move when pulled. If a rod is in a 0 state or has not been pulled, its gates are in their *resting position.*, as illustrated. If it is in a 1 state and has been pulled, its gates are in their *extended position*.

Basic Dynamics

Assume a rod having a series of 16 input locks and 16 output locks, with an alignment knob assembly between them occupying 2 nm. Given a lock length of 1.5 nm, the rod's total length will be 50 nm.

In any given cycle, first the input locks are set, then the displacement-source pulls the spring, then the output locks are probed. If one or more of the input locks is in a 0 state, then the rod beyond that lock cannot be pulled significantly: the probe knob for that lock lodges against the gate knob and the drive spring stretches to absorb the displacement. The gate segment remains in a resting position determined by the alignment knob and stop, together with a 0.5 nN net alignment force (2.5 nN applied by the anchor spring, minus 2.0 nN tension in the probe segment).

If all of the input locks are in the 1 state, then the displacement source can affect the entire rod. The stiff drive spring stretches only moderately, and hence transmits displacement.

As the probe segment tension passes 2.5 nN, it begins to move the alignment knob and gate segment. Eventually, the probe segment tension reaches 3.0 nN, the alignment knob is lodged against the second alignment stop by a net force of 0.5 nN, and the gate knob segment is in its extended position. The distance traveled by the alignment knob and gate segment (determined by gate displacements needed to effect logic state transitions) is 0.84 nm.

The probe segment tension varies by I nN, stretching it by about 1.1%, or 0.28 nm; since this sums with the alignment knob displacement, the drive spring must shift its end of the rod by about 1.12 nm. The gate segments tension remains constant, hence its length remains constant.

In the absence of detailed molecular mechanics calculations, rough approximations can indicate the magnitude of the dynamic variables in this motion. Assume a rod as described above, and assume that the drive spring imposes a smooth motion on its attachment point, amounting to a half-period of a sinusoid, taking a total of 50 ps. The peak speed is then:

$(\pi/2)[(1.12\times10^{-9})/(50\times10^{-12}\text{ s})] = 35\text{ m/s}.$

In the rigid-body approximation (which is pessimistic in assuming that all parts of the rod reach this maximum speed), the acceleration is $2.2\times10^{12}\text{ m/s}^2$, the accelerating force is 0.034 nN, and the peak kinetic energy (on a per-output-lock basis) is 6.0×10^{-22} J, or about 0.14 kT at 300 K.

Since the speed of sound in the rod is about 17,000 m/s, mechanical signals will propagate from the displacement source to the alignment knob in less than 1.5 ps, and to the far end in less than 3 ps.

These times are short compared to the 50 ps time for the overall motion, hence it is a fair approximation to neglect signal propagation time and the excitation of rod vibrations by the displacement source (collisions with the alignment stops are another matter.

The maximum stretching of the rod due to signal delay is about 0.01 nm, and the associated elastic energy per output lock is about 0.001 kT.

FRICTION, COMPRESSION, AND ENERGY DISSIPATION

Static Friction

In conventional bearings, surface contacts cause substantial friction, both static and sliding. But in these bearings, surface asperities cause wear and other effects quite unlike those found in the motion of logic rods along their tracks.

The chief rod-track interactions result from van der Waals and electrostatic potentials; the essential smoothness of these potentials makes a contactfree electromagnetic bearing a better macroscopic analogy than a sliding bearing.

Since friction- effects will be dominated by the remaining bumpiness, the steeply varying van der Waals potentials are the main concern.

Figure shows the variation in van der Waals potential as one carbon atom moves past another along a series of straight lines with different distances of closest approach. Figure also shows the variation in van der Waals potential as one carbon atom moves past a row of other atoms, at a fixed closest-approach distance of 0.3 nm, but with varying atomic spacings in the row.

The dramatic smoothing of the potential as the effective row spacing approaches 0.1 nm can be explained in terms of a cutoff in the spatial frequency of the single-atom potential fields along the lines of travel.

Figure shows how the effect of short single-row spacings may be achieved using multiple rows.

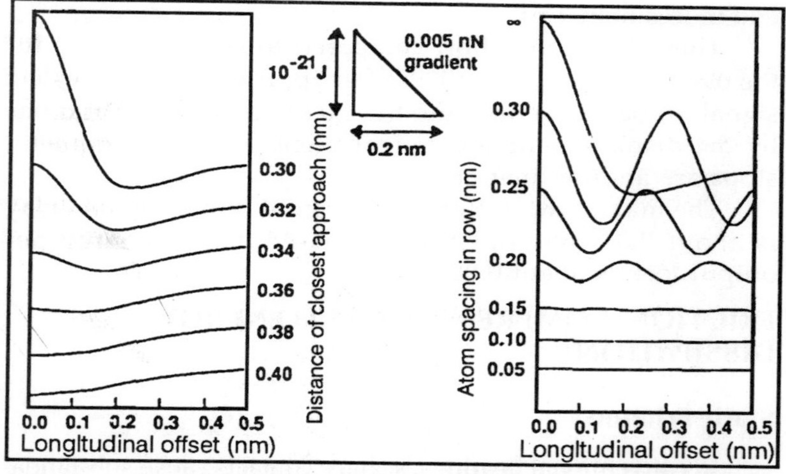

Van der Waals interaction potentials between carbon atoms. Left, between two atoms, one fixed and one moving along lines with varying distances of closest approach. Right, between a similarly moving atom and rows of atoms, all with a 0.30 nm closest-approach distance, but with varying in-row spacings. Similar results are found at closer approach distances.

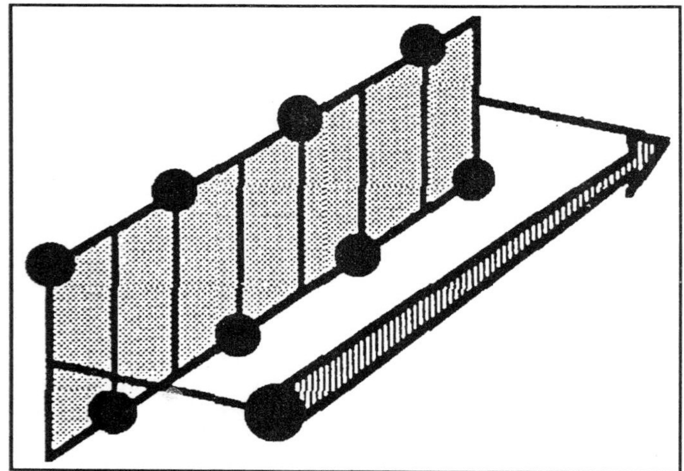

Fig. Effective Spacing of Stationary atoms Equals one Half their Minimum Physical Spacing

Each row of atoms in a track has a spacing on the order of 0.24 nm. By staggering the rows properly on both sides of the rod, this can give the effect of a single row with a spacing of about 0.06 nm. The resulting potential is quite smooth, with residual longitudinal forces (for a uniform track) of less than 0.0001 nN/atom. Roughly equal numbers of atoms in the rod will experience forces of positive and negative sign, hence these forces to a first approximation will cancel.

This longitudinally smooth potential can be modified by rendering the track nonuniform: local changes in track spacing can create smooth, local potential barriers and wells. These nonunifor- can be created by building nonuniformities into the covalent structure supporting the rows of track atoms. Interactions with the bottom surface of the knob-base rings offer further opportunities to introduce controlled nonuniformities in potential.

Knobs will also interact with non-track portions of the matrix and with other knobs. By choosing knob-wall spacings near the smooth, 0.4 nm contour of Figure and shaping the matrix properly, the many interactions between knob and wall atoms can be made to sum to a smooth potential. Longitudinal potential variations will remain, caused by electrostatic forces, crosschannels, and knob-knob interactions. Track nonuniformities can be designed to cancel these variations; in general, they need do so only when energies are summed over several adjacent locks. This may be called "tuning the track."

These considerations suggest that the maximum longitudinal force on the rod resulting from matrix and knob-knob interactions can be kept well under 0.1 nN. This force, which roughly corresponds to static friction, is acceptably low.

Sliding Friction

Unlike static friction, sliding friction dissipates energy. The two main mechanisms are acoustic radiation and phonon scattering (though "phonon" may be a misnomer where the inhomogene- matrix is concerned).

Acoustic radiation seems a minor loss. An elastic disturbance moving smoothly at subsonic speeds will radiate no sound; a knob moving along a track matches this description fairly well.

Only oscillating force components will cause radiation. These forces are small (for the same reason that the track is smooth), and have frequencies on the order of (35 m/s)/(0.3 nm) » 10^{11} Hz. Since the speed of sound in the matrix will be on the order of 10,000 m/s, the entire rod length will be in the near field of any given disturbance. Under these circumstances, oscillating forces on an atomic scale will chiefly store and recover elastic energy, not radiate it; losses should be low.

A crude estimate of the energy dissipated through phonon scattering (per lock, per operation) is $P \times f \times A \times Dt \times (v/c)^2$, where P = the power density of phonons in the matrix (about 5×10^{11} W/m^2 at 300 K), f = a correction factor reflecting the limited coupling between longitudinal knob motion and matrix motions (0.1 or less), A = the area of the knobs (10^{-18} m^2), Dt = the time of motion (50 ps), v = the speed of motion (35 m/s), and c = the speed of sound in the matrix (10,000 m/s).

The estimated loss per lock per operation is then $3 \times 10^{-}$

[23] J, or about 0.007 kT; a more accurate estimate would likely be lower.

Thermodynamic Effects

Knobs undergo thermal vibrations within elastic constraints resulting from rod tension, knob bending stiffness, and van der Waals interactions with the track and other nearby structures. These motions are physically analogous to those of a gas molecule in a container. For various rocking, bending, and bobbing modes, the confining forces can be approximated by springs with force constants on the order of 10 Tl/a, resulting in typical thermal excursions of roughly 0.03 nm.

When a probe knob presses against a closed gate knob, the resulting van der Waals interactions increase one constraining spring constant on the knobs by roughly fourfold, reducing typical excursions on that axis to about 0.015 nm. If we regard the moving knob as a moving gas parficle, that gas has undergone compression by a factor of 0.5. A variety of other events in the rod logic cycle will cause varying (generally lesser) degrees of compression. These events include moving a probe knob past an open gate knob and moving a gate knob toward a gate stop.

In an isothermal compression from volume V_1 to V_2, the work consumed would be $\ln(V_2/V_1)$kT. For a ratio of 0.5, this has a magnitude of 0.7 kT. In an isothermal cycle of compression and expansion, the net work consumed would be zero, and the resulting time-average forces would be like those of a lossless nonlinear spring. In this approximation, compression effects simply add another position-dependent potential energy function in the description of rod motion, one which can be smoothed by track tuning.

Deviations from isothermal compression cause what seem to be the dominant energy losses in the rod logic system. These deviations may be described in terms of the fractional change in volume per thermal coupling time. The resulting losses may be estimated by assuming that, during compression, the temperature is always that of an isothermal process that became adiabatic one thermal coupling time before the present moment.

Estimates of this coupling time can be made in different ways with similar results. If we assume coupling through collisions with walls during compression, then we get tens of collisions during compression (based on geometric considerations and the observation that the thermal speed is roughly 20 times the rod speed at which most compression occurs). Alternatively, we can note that knob vibrational modes couple strongly to transverse rod modes, which couple strongly to the thermal bath of the matrix. The pre-compression vibrational period of a knob will be about 0.5 ps, and frequency will increase during compression. Therefore, we expect some tens of cycles during a compression event lasting 5 ps. Assuming substantial equilibration of knob temperature per collision or vibration, the thermal coupling time will be on the order of one tenth the compression time, making compression significantly nonisothermal.

Assume a 10% change in volume during a thermal coupling time. Knob temperature (and work done) will then be roughly 5% too high during compression, and 5% too low during expansion, causing a loss per cycle of about 0.1 $\ln(V_2/V_1)kT$, or 0.07 kT for a compression ratio of 0.5.

Overall compression losses per lock per cycle may be four times this value, or about 0.3 kT. This estimate is very rough, but should be of the right order.

Overall Energy Dissipation

A further loss results from the dissipation of kinetic energy in the rod when a moving alignment knob strikes its alignment stops; this loss (taken as roughly one fourth the peak kinetic energy of the rod) amounts to about 0.04 kT per lock. Losses due to acoustic radiation and sliding friction seem negligible by comparison. The loss terms mentioned in this paragraph all scale as the square of the rod speed, or as the inverse square of the cycle time.

Compression losses seem dominant, totaling (very roughly) 0.3 kT per lock. These losses depend on the ratio of the knob thermal coupling time to the compression time—a time inversely proportional to rod speed. In the speed range of interest these losses should scale roughly as the speed of the rod, or as the inverse of the cycle time.

THERMAL NOISE

Variables

A	amplitude of mode (m)	L	distance, alignment knob to last gate (m)
b	rod bending stiffness (J-m)	M	modulus of rod (N)
c	track constraint (N/m^2)	n	mode number
d	displacement	N	number of atoms in model rod
D_e	minimum displacement for error (m)	P(d)	probability of displacement ≥ d
E	energy (J)	σ	standard deviation of knob position
$f_x(X)$	probability density function for X	σ_t^2	variance due to transverse modes (m^2)
F	fraction of energy in tension term	σ_l^2	variance due to longitudinal modes (m^2)
ϕ	net alignment force (N)	σ_r^2	variance due to rocking modes (m^2)
G	average of F^2 over all modes	T	temperature (K)
k	Boltzmann's constant (J/K)	τ	rod tension (N)
K	spring constant for knob rocking (N/m)		

Rod Displacement Model

As described in Section , the chief thermal noise problem is that a gate knob in a 0 state may move so as to allow a probe knob to pass, causing device-level errors. A reasonable goal is to keep this probability below one in 10^{12} (system-level errors can be made yet lower by means of standard techniques based on redundancy and error-correcting codes). Roughly speaking, this goal requires that the elastic energy associated with D_e exceed $\ln(10^{12})$ kT, or about 1.1×10^{-19} J at 300K. The following analysis will pursue a more detailed estimate based on explicit probability density functions for gate knob displacements.

The problem of thermal noise is most acute at the gate knob farthest from the alignment knob, since it stores the least elastic energy for a given displacement. Estimates of D_e must reflect the tendency of gate knobs to deform under load and shed probe knobs toward whichever side of the central sulfur atom they fail on; in near-error conditions, though, this is countered by elastic forces pulling the gate knob toward its proper position. Calculations suggest that a reasonable threshold displacement is one that carries the gate knob from its reference position (with a fluorine atom squarely in front of the probe knob) to a position in which the sulfur atom has moved at least 0.07 nm past the probe knob: this yields a D_e of about 0.34 nm. This tendency to shed probe knobs also has a favorable consequence. Probe knobs pressing against 0-state gate knobs push them toward their gate stops, increasing their resistance to displacement. In practical terms, a gate knob has only one chance per cycle to be misplaced.

The model used here considers displacements resulting from transverse vibrations, longitudinal vibrations, knob rocking, and shifting of the rod as a whole caused by motion of the alignment knob. Knob displacements are measured from a reference position, not from mean knob positions. The model is classical, and hence overestimates contributions from high-order vibrational modes. Description in terms of thermally excited vibrational modes here entails no assumption of regular sinusoidal motions; it can instead be regarded as simply an analytically tractible way to describe small, arbitrary rod deformations, driven by entropic forces. The bulk of the analysis ignores the effect of gate stops, but takes rough account of them as a correction at the end.

Rod Transverse Modes

In transverse thermal vibration, the average potential energy in a mode, 1/2 kT, is divided among rod-bending, rod-tension, and track-constraint terms. Track constraints are here modeled as linear elastic centering forces. Bending is treated in a continuum model, smoothing discrete bond bending stiffnesses . The mass distribution is treated as uniform, save in setting a cutoff in mode number; an N-atom chain supports 2N transverse modes, yielding a total energy of NkT. In this model, the average potential energy in a mode is:

$$E_n = \frac{A_n^2}{8}\left[bL\left(\frac{n\pi}{L}\right)^4 + \tau L\left(\frac{n\pi}{L}\right)^2 + cL\right],$$

and the fraction of this energy in the tension term is:

$$F_n = \left[\frac{b}{\tau}\left(\frac{n\pi}{L}\right)^2 + 1 + \frac{c}{\tau}\left(\frac{n\pi}{L}\right)^{-2}\right]^{-1}$$

Given $\tau = 2.5$ nN, c = 1 N/rn-atom = $7.6\text{x}10^9$ N/rn^2 (a low value), and b = $4.8\text{x}10^{-29}$ J-m, the average over all the modes is:

$$F_{average} = \sum_{n=1}^{N} \frac{F_n}{N} \approx 0.304 \text{(almost constant for large N)}.$$

The average total energy in the tension term is locally fitted by the linear expression:

$$E_T \approx NkT\left\{0.304 + 0.172\left[\left(\tau/2.5\times10^{-9}\right)^{-1}\right]\right\}.$$

Given the constant-tension boundary condition set by the anchor spring, the tension term in the average potential energy is reflected in a shortening of the average rod length by a distance E_T/t. Greater tension tends to straighten the rod, implying a finite modulus even without bond stretching. This yields the 1500 nN modulus correction parameter used in Section 3.3.

Deviations from this average energy cause variations in rod length, contributing a thermal noise term. Each mode may be treated as a harmonic oscillator with (classically) an exponential probability density function (PDF) for energy. This yields a PDF for the change in rod length due to the average potential energy in the tension term of mode n:

$$f_{d_{nt}}(d_{nt}) = \frac{2\tau}{F_n kT}\exp\left(-d_n \frac{2\tau}{F_n kT}\right), (d_n \geq 0).$$

(The average is taken because rod length responses are slow compared to transverse vibra tion times.) The variance in rod length due to mode n is then:

$$\sigma_{d_{nt}}^2 = F_n^2 \left(\frac{kT}{2\tau}\right)^2.$$

The variance in rod length due to fluctuations in the energy of all transverse modes is:

$$\sigma_1^2 = 2NG\left(\frac{kT}{2\tau}\right)^2 \approx 1.4 \times 10^{-15} L, \text{ with}:$$

$$G = \sum_{n=1}^{N} \frac{F_n^2}{N} \approx 0.132 \left(\text{G almost constant for large N}\right).$$

These individual exponential terms are small and numercus enough to make the PDF of their sum approximately gaussian in the range of interest.

Rod Longitudinal Modes

The energy of a longitudinal vibrational mode (treating the rod as clamped at one end and free at the other) is:

$$E_n = \frac{1}{2}\frac{\pi^2 M}{8L}(1+2n)^2 A_n.$$

The PDF for the energy will be exponential, yielding a Rayleigh PDF for the vibrational amplitude and a gaussian PDF for the instantaneous displacement of the free end of the rod:

$$f_{A_n}(A_n) = r_n^2 A_n \exp\left(-\frac{1}{2} r_n^2 A_n^2\right); \text{ with } r_n = \left[\frac{1}{kT}\frac{\pi^2 M}{8L}(1+2n)^2\right]^{1/2}$$

$$f_{d_n}(d_n) = \frac{r_n}{(2\pi)^{1/2}}\exp\left(-\frac{1}{2} r_n^2 d_n^2\right); \text{ yeilding } \sigma_{dn}^2 = \frac{1}{r_n^2}.$$

Summing over the modes yields the total variance in longitudinal vibrational displacement:

$$\sigma_1^2 = \sum_{n=0}^{\infty} \frac{1}{r_n^2} = \frac{8LkT}{\pi^2 M} \sum_{n=0}^{\infty} (1+2n)^{-2} = \frac{LkT}{M} = 4.6 \times 10^{-14} L$$

Gate Knob Rocking Modes

Gate knobs can rock in the longitudinal direction, but at the cost of storing potential energy in bond bending and rod deflection. A model that accounts for rod tension and for bond-bending at both the ring attachment points and the first atoms in the inter-knob segments (using parameters from and) yields an estimated restoring force of *9.5* N/rn at the effective height of the gate knob. (This estimate neglects favorable track constraint forces.) This implies a gaussian PDF with variance:

$$\sigma_1^2 = \frac{kT}{K} \approx 4.4 \times 10^{-22} \, m^2.$$

Alignment Knob Displacements

Movement of the alignment knob away from its stops will displace the entire rod. This displacement is directly proportional to the increment in energy and inversely proportional to the alignment tension, resulting in a PDF:

$$f_{d_e}(d_e) = \frac{\phi}{kT} \exp\left(-\frac{\phi}{kT} d_e \right).$$

Design considerations make it desirable to minimize fsubject to the error-rate constraint. This results in an exponential term too large to treat by simply adding its variance to those above and treating the sum as gaussian.

Overall Displacement Probabilities

Taking L as 24 nm, the sum of the variances of the gaussian terms above is:

$$\sigma = \left(\sigma_t^2 + \sigma_1^2 + \sigma_r^2 \right)^{1/2} \approx 4.0 \times 10^{-11} m,$$

and the associated PDF is:

$$f_{d_g}(d) = \frac{1}{\sigma (2\pi)^{1/2}} \exp\left[-\frac{1}{2} \left(\frac{d_g}{\sigma} \right)^2 \right].$$

The effect of gate stops, which constrain several of these positional displacements to be predominantly in a single direction, may be conservatively approximated by doubling the probability density in the adverse tail of the gaussian term. Including this factor of two, the probability of finding the rod with a displacement greater than D_e is:

With $\phi = 0.5$ nN and the above values for other parameters, P_e equals 5×10^{-13}, meeting the stated condition on errors. Satisfaction of the chosen error condition is a matter of design, not of chance. If better models, changed parameters, or a more stringent error condition should adversely affect this result, changes in parameters such as rod tension, alignment force, and gate knob length could be made, resulting in a generally similar design that again met the error condition.

Molecular-scale mechanical logic elements with fan-in and fan-out ratios of 16 or more and error rates of less than 10,12 appear feasible. These elements-rods analogous to conducting paths and locks analogous to transistors-are suited to the implementation of conventional computer architectures and meet the relevent criteria for good computer devices.

Analysis reveals thermal noise as a major but acceptable constraint on logic element design. Operating delays for elements having 16 inputs and outputs are on the order of 50 ps, a time consistent with the implementation of mechanical nanocomputers having gigahertz clock rates. Rod logic elements have linear dimensions on the order of one ten-thousandth (and volumes on the order of one trillionth) of those of commercial VLSI circuit elements of the recent past.

The volume per lock in a regular array is about 4.1 cubic nanometers, neglecting overhead for components roughly analogous to power, ground, and clock signal paths. The energy dissipated during lock operation is speed dependent; at the speed assumed here, it appears to be on the order of 0.4 kT per lock.

The design approach chosen is robust, in that the feasibility of devices of the sort described does not depend on the precise values of uncertain physical effects or parameters. A valuable contribution to the study of mechanical nanocomputers would be the use of existing molecular mechanics programs to model the dynamics of rods and locks in greater detail, to verify their gross behaviour and to produce better estimates of energy losses.

By demonstrating a molecular mechanical technology able to implement reliable logic systems, this work shows the possibility of molecular machines able reliably to execute complex patterns of motion. This lends further support to the concept of molecular assemblers able to build structures to complex, atomic specifications, and to the feasibility of a wide range of technologies relying on computer-directed machinery with parts of molecular dimensions.

Chapter 5

Molecular-Scale Computer Device

In 1985, Keyes argued that the silicon transistor would not be easily replaced as the building block in digital circuits. Although faster switching technologies have been introduced, none possess the wide array of desirable features to match silicon: gain, voltage restoration, noise reduction, tolerance to variability in the manufacturing process, isolation of inputs from outputs, excellent fan-out, the ability to invert a signal, tolerance to the high demands of a system environment, and a huge body of research and financial commitment to their use and manufacture.

Since that time, advances in CMOS technology have increased the speed and decreased the size of transistors at a phenomenal rate; recent research has produced 30-nm gate CMOS transistors that switch in less than 2 ps. Impressive as these results have been, in the near future further increases in the performance of silicon-based, lithographically manufactured transistors will be difficult to achieve. Never before has there been so much doubt in the industry about how three-generation-removed advances will be accomplished. There are several major reasons for this.

First, the small line widths necessary for next-generation lithography require the use of increasingly shorter-wavelength light, which introduces a host of problems that are currently being addressed. In addition, as the number of atoms that constitutes a device decreases, manufacturing variability of even a few atom widths can become significant and lead to defects. More important, however, is the economic barrier to commercial nanometer-scale lithography.

New fabrication facilities orders of magnitude more expensive than present ones will be needed to produce chips with the required densities while maintaining acceptably low defect rates. The increasing cost of chip masks, which must be manufactured to single-atom tolerances, precludes commercially viable development of new chips except for the highest-volume ICs. It is entirely possible that further reduction of transistor size will be halted by economic rather than technological factors.

This economic downside is a direct consequence of the precision required for deep-submicron lithographic fabrication. Using lithography, construction of devices and their assembly into circuits occur at the same time. Keyes pointed out that this is a great advantage of silicon in that mass fabrication of transistors is extremely inexpensive per transistor. However, as Keyes also highlighted, it produces a set of constraints: each element in the system must be highly reliable, and devices cannot be tested, adjusted, or repaired after manufacture. These constraints force the design of a custom circuit to be tightly integrated with manufacture, since no additional complexity can be introduced into the circuit afterwards.

Lithography is perfectly tuned to deal with these constraints, but at small scales the precision required to obey these constraints becomes the obstacle to further improvement. Any technology hoping to displace lithographically produced CMOS integrated circuits must overcome such obstacles while sacrificing neither performance nor the low per-unit cost made possible by mass production.

ELECTRONIC NANOTECHNOLOGY

Chemically assembled electronic nanotechnology (CAEN) holds promise as a technology to overcome these obstacles. CAEN takes advantage of chemical synthesis techniques to construct molecularsized circuit elements such as resistors, transistors, diodes, resonant tunneling diodes (RTDs) , and reconfigurable switches. Chemical synthesis can produce enormous quantities (moles) of identical devices cheaply, or devices can be grown in situ. In all cases, the fabricated devices are only a few nanometers in size and exploit quantum mechanical properties to control voltage and current levels across the terminals of the device. Most of these molecules are smaller than the physical feature-size limit that can be produced using silicon.

Additionally, tiny amounts of current are needed to operate these devices, which should result in incredibly low power consumption in devices created from them. Molecular-scale devices cannot by themselves overcome the constraints that will prove limiting for CMOS unless a suitable, low-cost manufacturing technique can be devised for composing them into circuits.

For devices created separately from the circuit assembly process, localization and connection of the devices using a lithographic-like process will be very difficult, and other means must be sought. Connecting the nanoscale components one at a time will be prohibitively expensive due to their small size and the large number of components to be connected. Instead, economic viability requires that circuits be created and connected through self-assembly and self-alignment. Several groups have recently demonstrated CAEN devices that are self-assembled or self-aligned, or both.

Advances have also been made in creating wires out of single-wall carbon nanotubes and aligning them on a silicon substrate. More recently, selective destruction has been used to create desired carbon nanotube structures. In addition, metallic nanowires, which can scale down to 5nm and can include embedded devices or device coatings, have been fabricated.

Self-assembly of devices to form circuits imposes an entirely different set of constraints from lithography. Because self-assembly will be an imprecise, thermodynamically controlled process, we can no longer expect to produce working devices in specified locations with the high reliablity of CMOS lithography. This implies that the devices must be testable, adjustable, and possibly repairable after manufacture, and also that some defects must be tolerated.

Moreover, self-assembly can produce only a limited range of circuit complexity while remaining manageable. Thus the manufacturing process must be decoupled from the production of the final circuit. Any deterministic aperiodic circuit will have to be created through reconfiguration after the device is fabricated. Reconfigurability will therefore be a key component of any successful nanoscale architecture. The regular structures amenable to creation via self-assembly are ideally suited to creating a reconfigurable computing fabric. Reconfigurable computing uses large arrays of uniform, programmable logic elements and interconnects, which can be configured using electrical signals, to perform required computations.

Reconfigurability has been used for years to implement new circuits after manufacture, as in commerical and research Field Programmable Gate Arrays. It has also been used to incorporate defective circuit elements into a working, high-performance computer, the Teramac, by discovering the defective circuit elements after fabrication. Molecules that can act as reconfigurable switches in series with diodes have already been discovered. These switch/diode pairs have a great advantage over their CMOS counterparts: configuration information is stored in the device itself, rather than in separate static RAM cells, greatly decreasing the area required for each circuit element. A single CAEN switch-diode pair will require 800 nm^2_ as opposed to 100,000 nm^2 _ for a single laid-out CMOS transistor.

A simple CMOS logic gate or static memory cell requires several transistors, separate p- and n-wells, and wiring interconnects, resulting in a factor of more than one hundred difference in density between CAEN and CMOS. While one switch/diode does not implement the same functionality as a transistor, clearly the size differences give CAEN technology an edge. The use of self-organization as the dominanant means of circuit assembly imposes the most severe limitation on nanoscale architectures: precise device alignment will be very difficult to achieve. Chemical self-assembly, as a stochastic process, will produce precise alignment of structures only rarely, and manipulation of single nanoscale structures to construct large scale circuits is impractical at best. Therefore, the most reliable way to make device connections will be to generate them at the intersection of two wires coated with the appropriate molecular devices.

This alleviates the need for precise or end-to-end connections, as devices are incorporated into the circuit out of sheer topological necessity. However, connecting together more than two devices at a time, or devices with more than two terminals, will likely prove extremely difficult to do *en masse* at the nanometer scale. Three-terminal devices such as molecular analogs of transistors will not be practical, since ultimately the terminals of these devices will need to be connected in precisely defined patterns to produce properly functioning circuits. In practical, near-term nanocomputer designs, the active components will be two-terminal devices such as diodes, configurable switches, and molecular RTDs.

A Plausible Nanoarchitecture

Having enumerated the constraints imposed by CAEN, we present the highlights of a plausible architecture and fabrication process for constructing a molecular integrated circuit. The process, a hybrid of molecular self-assembly and lithographically produced aperiodic structure, is inexpensive yet creates structures out of nanoscale components. At the nanoscale, all subprocesses are self-directed; only at the micron scale do we allow deterministic operations. The process is hierarchical, proceeding from basic components (e.g. wires and switches), through self-assembled arrays of components, to complete systems. Since practical transistors will be difficult to achieve at the nanoscale, computation is performed using diode/resistor-based threshold logic.

In the first fabrication step, wires of different types are constructed through chemical selfassembly in bulk. Some of these wires may be functionalized, either by coating their exterior or by including inline devices along the axis of the wire. The next step aligns groups of wires into rafts of parallel wires using flow techniques. Two rafts can be combined to form a two-dimensional grid using the same flow techniques. The active devices are created wherever two appropriately functionalized wires intersect.

By choosing the molecules that coat the wires appropriately, we can make the intersection points ohmic contacts, diodes, configurable switches, or other devices. No precise alignment is required to create the devices; they occur wherever the rafts of wires intersect. The resulting grids will be on the order of a few microns in size.

A separate process will create a silicon-based die using standard lithography. Note, however, that this lithographic process does not have to be "state-of-the-art" since the CMOS components are few and far apart. The circuits on this die will provide power, clock lines, an I/O interface, and support logic for the grids of devices. The die will contain "holes" in which the grids are placed, aligned, and connected with the wires on the die. To create useful aperiodic circuits using CAEN, reconfigurable devices will be incorporated into the grids. The main active component at the intersections of the grid will be a reconfigurable molecular diode (or a switch in series with a diode). The diode can be configured "off", in which case the it acts like an open circuit, or "on," in which case it acts like a diode.

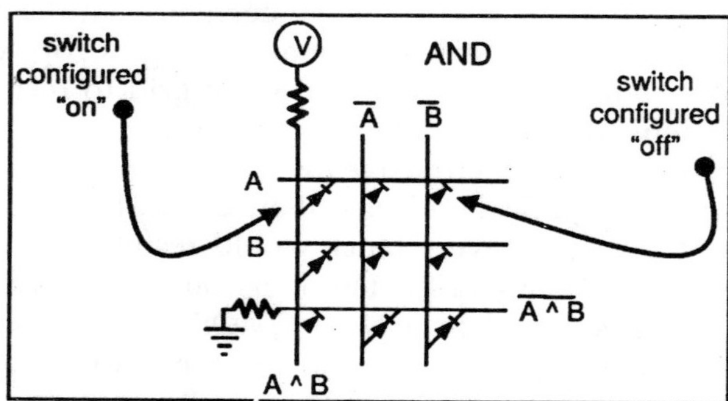

Fig. A two-input AND Gate Implemented in a Reconfigurable CAEN Grid.

Figure shows how a 3x3 grid might be configured to implement an AND gate. Four of the nine reconfigurable diodes are configured "on." Since inverters cannot be constructed using diodes and resistors, CMOS inverters will be required at the input to the CAEN-based logic, which must compute both the desired output and its complement at every stage in the computation. The nondeterministic nature of self-assembly will give rise to high defect densities. Instead of trying to eliminate these defects completely with manufacturing techniques, we perform postfabrication steps that allow the chip to work in spite of its defects.

A natural method of handling the defects, first used in the Teramac, is to exploit the reconfigurable nature of the device. First, the device is configured for self-diagnosis. The result of the self-diagnosis phase is a defect map. Then the defect map is used to configure the desired functionality around the defects. Reconfigurability is thus integral to the operation of a CAEN-based architecture.

Signal RestorationWithout Transistors

The architecture described above satsifies the key constraints of nanoscale technology: it uses no deterministic operations at the nanoscale, and it uses only two-terminal nanoscale devices. To build an operational circuit, however, requires signal restoration between nanoscale wire grids. Connections to ground will remove current from the circuit, and every diode will cause a small voltage drop. CMOS transistors could be used between every grid to restore the signal to defined voltage and current levels.

However, this is unattractive for two reasons. First, CMOS transistors are significantly larger than CAEN-based devices; requiring their use at every block would result in an overall decrease in the density of the fabric. Second, if a low-current CAEN signal is forced to drive a CMOS signal-restoration transistor, the relatively large capacitance of the CMOS transistor will decrease the circuit's overall speed.

Here we describe an alternative: a molecular latch, constructed entirely from molecular-scale devices, that can perform signal restoration using power from the clock to provide gain. This idea is motivated by early work on tunnel diodes and more recently on compound semiconductor resonant tunneling diodes.

The molecular latch provides the important properties of signal restoration, I/O isolation, and noise immunity.

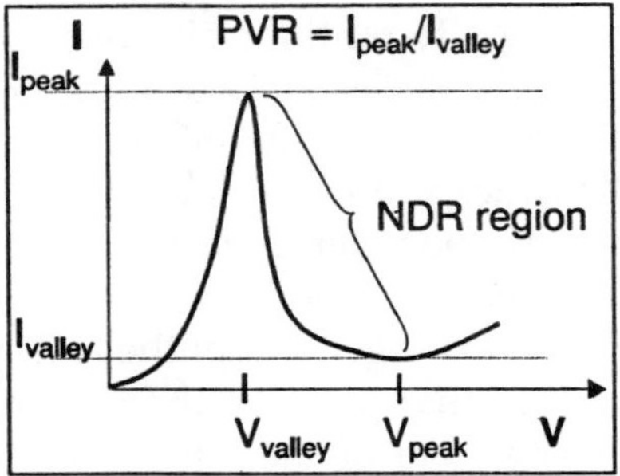

Fig. I-V Curve for a "Typical" RTD device.

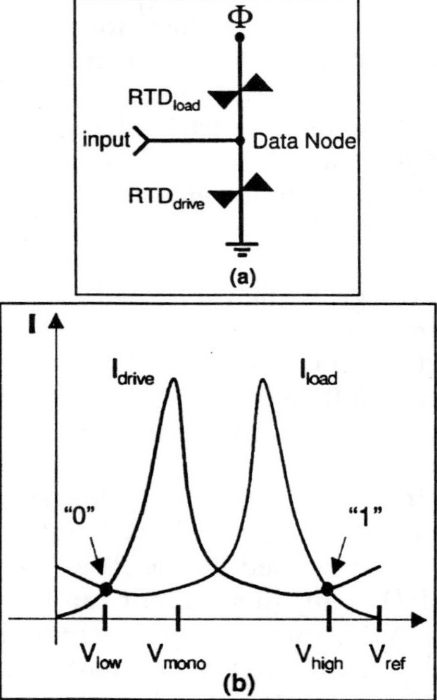

Fig. (a) Basic Molecular Latch Configuration. (b) Load-Line Diagram for Molecular Latch.

Molecular Latch Operation

The molecular latch is constructed from a pair of molecular resonant tunneling diodes (RTDs). Figure shows an I-V curve of a representative RTD. The key feature of an RTD is a region of negative differential resistance (NDR), where the tunneling current falls as the voltage increases. The ratio of the current at the beginning of the NDR region to the current at the end is called the peak-to-valley ratio (PVR). Molecular RTDs that operate at room temperature with PVRs of 10 or more have already been realized.

Figure shows the arrangement of the RTDs in a molecular latch and a load-line diagram for the molecular latch at a voltage $V_{ÀÛÜ\ ref}$. The state of the latch is determined by the voltage at the node between the two RTDs. The load line diagram shows two stable states, V_{low} and V_{high}, and a third metastable state. Small voltage fluctuations in the metastable state will push the circuit into one or the other of the stable states. The low state represents binary "0" and the high state binary "1." The state of the latch is changed by temporarily disrupting and restoring this bistable equilibrium state. The bias voltage is temporarily lowered to V_{mono}, causing a shift of the load line to the left such that the circuit has only one stable state.

The bias voltage is then returned to V_{ref}. During the evolution from the monostable to bistable state (the monostable-bistable transition, or MBT), current introduced to the data node will flow through the drive RTD. If the total charge introduced by this current during the transition exceeds a threshold value, the circuit will switch to the high state; otherwise the circuit will switch to the low state.

The amount of current necessary to force the latch into the high state is determined by the PVR of the RTD. For a well-matched RTD pair, this current can be very small. A higher PVR requires more current to set the high state, but provides greater stability against current variation. For the circuit to function correctly, the drive RTD must be more resistive than the load RTD, i.e., R_{ser} in Figure must be larger for RTDdrive. RTD latch technology has already been exploited in constructing a GaAsFET logic family by Mathews et al..

They describe a series of logic gates constructed from RTD latches, FETs, and saturated resistors that display high switching speeds and low power dissipation. In CAEN, the RTD latches are used for a different purpose: they provide voltage buffering by storing the state of previous computation for use in later computation, and signal restoration to either the '0' or '1' voltage with each latch.

All computation is performed by diode-resistor logic and the latch restores the signal for a later computational stage. Work in the 1960s with tunnel diodes and threshold logic was hampered by high interdevice manufacturing variability ; the molecular RTDs discovered thus far have higher PVRs than semiconductor-based devices, which, as we explain below, may alleviate some of these problems.

Molecular Latch Simulations: Restoration and Reliability

Since molecular-scale RTD latches have yet to be realized, we conducted simulations of their behaviour using SPICE. We used a similar device model as Mathews et al. for each RTD The device model is depicted in Figure 4a. The RTD is modeled as a series resistance (R_{ser}), with a parallel capacitance (C_{RTD}) and voltage-dependent current source (L(V)). This meshes well with the proposed construction of molecular-scale RTDs, which is accomplished by connecting two segments of molecular-scale metal wire with a number of identical RTD molecules (See Figure 4b). Based on the estimated length and dielectric of the molecules and diameter of the molecular wires, we estimate a capacitance of 2.7 aF for each RTD.

We constructed model equations for the voltage-dependent current source to mimic the I-V curves reported by Reed and Tour at 190K. To explore the RTD parameter space, we also constructed models for several different "ideal" molecular RTDs in which the V_{peak} and V_{valley} locations and PVR were varied.

Figure shows the results of a simple RTD simulation using an idealized molecule as the basis of the RTD latch circuit shown in Figure. The clock signal is shaped to provide a monostable-bistable transition (MBT) in every cycle and to maintain the bistable state for the remainder of the cycle. Note that the cycle time was deliberately made longer than required to ensure equilibrium before the next cycle. The figure shows that the voltage is restored to the levels predicted by the RTD load line and the choice of V_{ref}.

In contrast to the "ideal" molecule, the relatively high V_{peak} of the Reed-Tour molecule would produce narrowly spaced '0' and '1' voltages for any reasonable choice of V_{ref}. The molecular RTD latch becomes more practical as V_{peak} becomes lower; the closer V_{peak} is to 0V, the larger the separation of low and high signals that can be produced at a realistically achievable V_{ref}. To determine the envelope of stability for the RTD, we simulated the behaviour of this simple circuit while varying the 'low' and 'high' input voltage, the input current, and the relative sizes of the load and drive RTDs. In simulation, simple RTD latches in isolation appear to be stable over the range of variability likely to be encountered in their synthesis. Stability has proved a significant

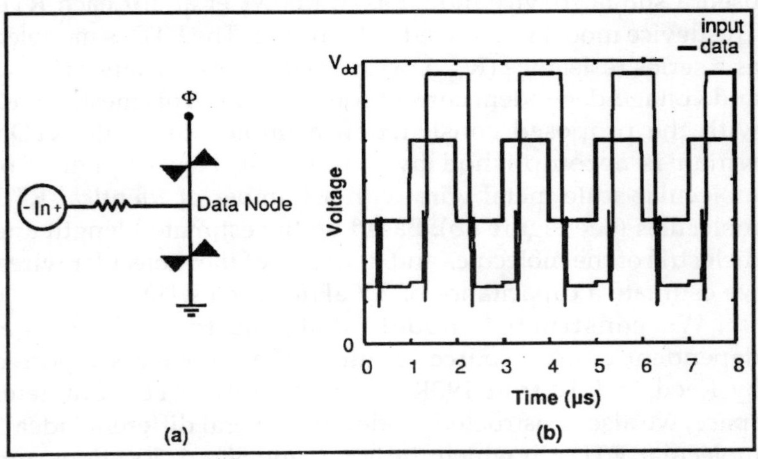

Fig. (a) Simple molecular latch circuit used in testing.
(b) Results of single Stability Experiment. Note voltage "gain." and Memory effect.

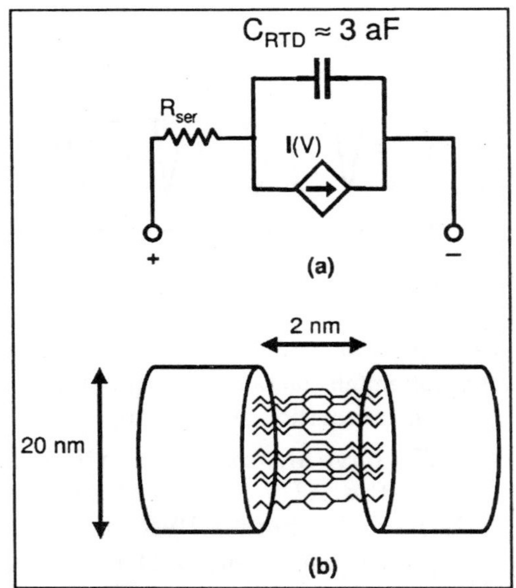

Fig. (a) Device Model for RTD. (b) Possible construction technique for molecular RTD as an inline device (not to scale).

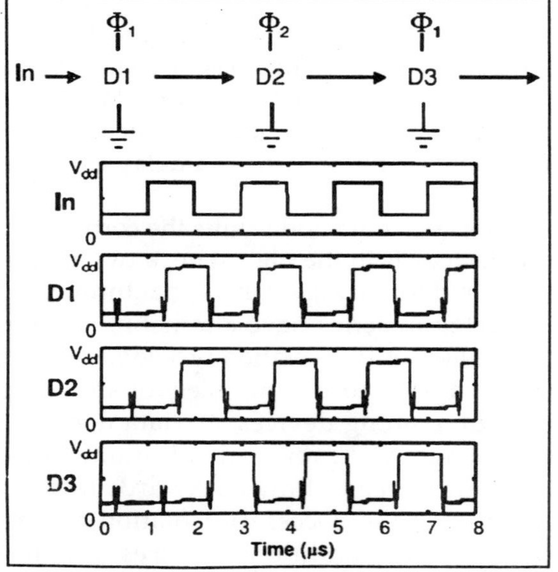

Fig. Simple 3-stage delay Circuit using Molecular latch

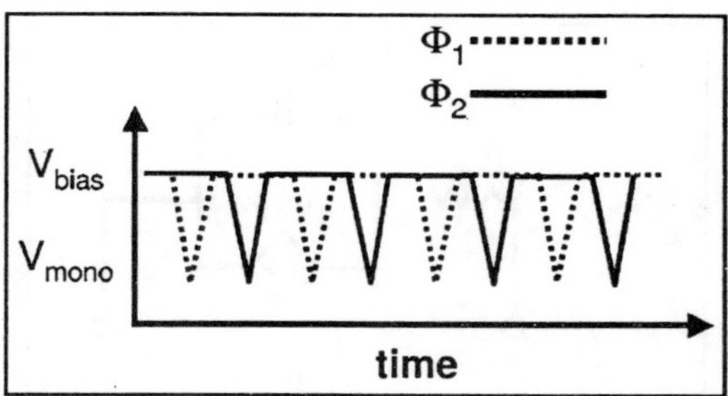

Fig. Two-phase clocking scheme used for series latches Stumbling block to the Application of Semiconductor RTDs, as device Variability often falls outside the required range for Incorporation into Circuits.

Molecular RTDs have an advantage in that all molecules incorporated into the RTD are identical. In addition, molecular RTDs have far bigger PVRs than conventional RTDs, which improves the stability of latches constructed with them.

Combinations of latches: I/O Isolation

In molecular circuits, latches are used to buffer and condition the output of a diode-resistor combinational circuit so that it can serve as input to the next one. We therefore need to consider the interactions of multiple latches. The simplest circuit using more than one latch is a delay circuit, as shown in Figure.

Because the latch data node emits the correct voltage only when the latch clock is in the high state, a two-phase clocking scheme is required to propagate the signal through the circuit. Latches connected in series have alternating clock phases. It is clear that the lack of I/O isolation provided by transistors is problematic for a latch consisting solely of an RTD pair.

Without intervening devices, all data nodes must exist at the same voltage, and even with additional linear circuit elements current is free to flow backwards to set upstream latches. To provide the necessary isolation characteristics, we incorporate several additional devices into the latch, as shown in Figure.

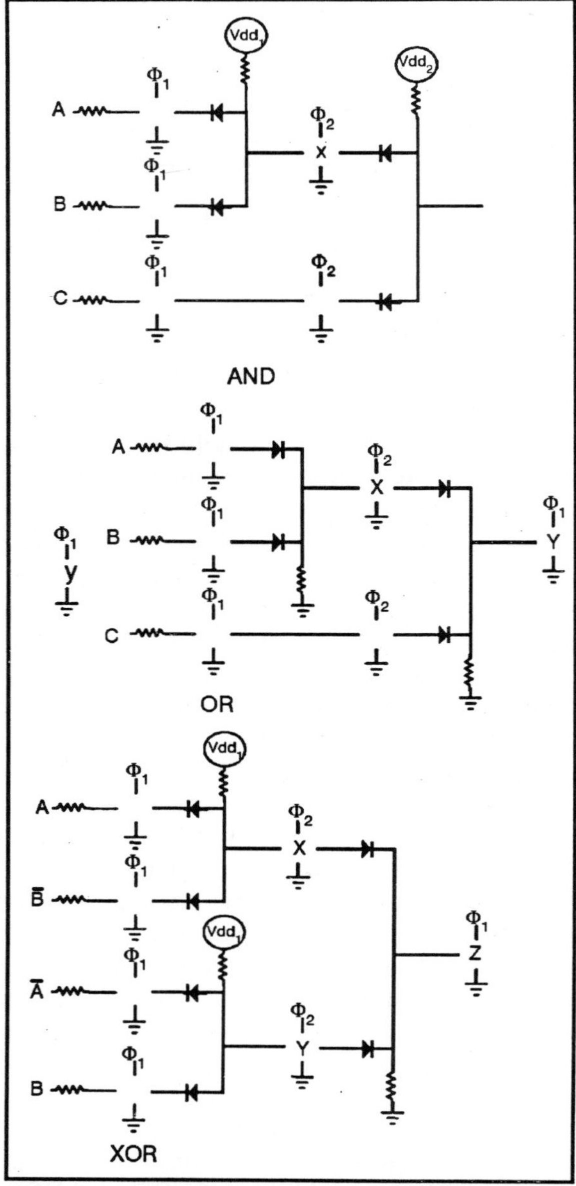

Fig. AND, OR, XOR Circuits using diode-resistor logic and Intermediate Molecular Latches set an upstream latch.

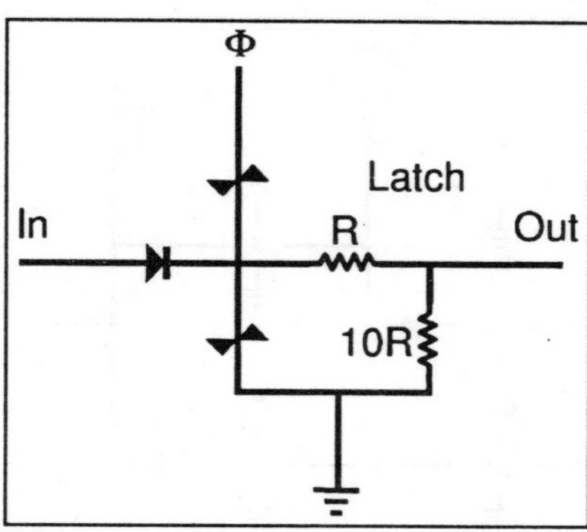

Fig. The Complete Molecular latch.

A molecular-scale diode is added to prevent current from a downstream latch from flowing to without the diode, setting a downstream latch will prevent the immediately upstream latch from resetting properly. Molecular diodes with very low voltage drops and reverse current flow have already been realized. The resistor immediately following the latch data forces enough current into RTD_{drive} to allow it to switch high, while allowing enough current into the next latch to determine its value. Without the resistor, latches in series would act as a current divider, and insufficient current would flow through each RTD_{drive} to set it to the '1' state.

To understand the rationale for the resistor to ground, consider the delay latch shown in Figure 6 and the circuit state (In=Low, D1=Low, D2=High, D3=High) at the instant before the Φ_2 clock cycle. As the Φ_2 clock voltage is lowered to V_{mono}, the diodes in both D2 and D3 are reverse biased, preventing any current discharge from the capacitance of latch D2. Since discharge is necessary for the latch to enter the low state, we provide a path to ground through a large resistor. The value of this resistor must be large to prevent current loss that would cause latch-setting failure.

However, the resistor should also be as small as possible in order to minimize the latch reset time. We simulated several simple circuits using the above latch configuration, including the delay circuit in Figure and the AND, OR, and XOR circuits shown in Figure. The circuits calculate the correct values, taking into account the delay required to traverse intervening latches. Circuits with more logic levels were also simulated successfully. One additional complication resulting from a lack of I/O isolation was discovered for the AND

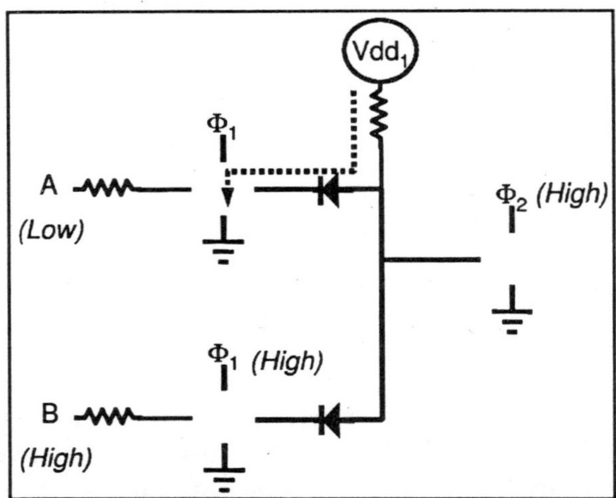

Fig. Lack of I/O isolation in AND Circuit Provides path from V_{dd1} to ground

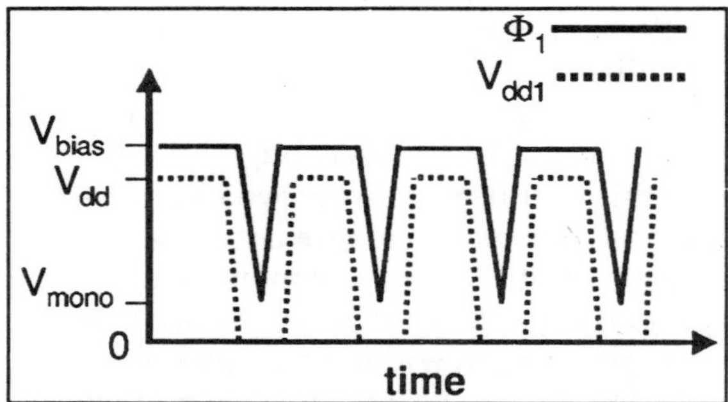

Fig. Clocking scheme used for V_{dd}. circuit.

Consider the circuit state (A=low, B=high, Y=high, Z=high) immediately beforeΦ_1 begins its cycle (Figure). The diode of Z is reverse biased, so the current in V_{dd1} will flow through the AND diodes into the latches and to ground, preventing them from being reset. The latch cannot be protected by a diode to prevent this, because doing so leaves no path to sink current from the pullup voltage. Without a path to ground, the pull-up would would cause Z to be set high incorrectly.

To counteract this, we introduce a clocking scheme for the pull-up voltage as well, as shown in Figure. The pull-up voltage is temporarily brought to ground during the MBT of the preceding latch. This removes the forward influence and allows the latches to be set properly. Unfortunately, in a real circuit this will likely have deleterious consequences for the RC constant.

Latch Summary

In simulation, the molecular latch is capable of restoring voltage levels to the two stable values and providing current input, as determined by the IV curve of the underlying RTD. It appears to be stable against variability brought about by manufacture, as long as the drive RTD can be reliably made more resistive than the load RTD. Because of the relatively low current required to switch states, one molecular latch can drive several other latches, i.e., latches can have moderate fan-out. In combination with the clocking scheme described above, the latch also provides the necessary I/O isolation to ensure proper calculation. Some caveats are necessary, however. Because the RTD devices discovered thus far are ex-tremely resistive, the RC constant for circuits incorporating them is very long (on the order of hundreds of nanoseconds).

Clearly, this will need to improved before these devices become practical. Also, the relative sizes of the various resistors must be carefully controlled during manufacture to ensure that they are properly matched; this will allow more devices to attain proper I/O isolation without adversely affecting the time constants for the circuit. Lastly, the use of a clocked pullup voltage may result in additional complications for the design of the CMOS support circuitry. However, the similarity in waveforms between the latch clock and V_{dd} clock may alleviate some of these problems.

Any architecture based on CAEN must overcome the constraints associated with directed assembly of nanometer-scale components and yet exploit the advantages of molecular electronics. Because the fabrication process is fundamentally nondeterministic and prone to introducing defects, the device must be reconfigurable and amenable to self-testing. This will allow the characteristics and defects of each device to be discovered, support the creation of circuits that avoid the defects, and fully exploit the available resources. For the foreseeable future, the fabrication processes will produce only simple, regular geometries.

Therefore, proposed architectures should be built out of simple, two-dimensional, homogeneous structures. Rather than fabricating complex circuits, CAEN will rely on the reconfigurability of regular fabrics to implement arbitrary functions after manufacture.

The construction process also needs to be parallel; heterogeneity should be introduced only at a lithographic scale. The resulting devices will be much like today's FPGAs in that they can be configured (and reconfigured) to implement any circuit in the field, but they will have several orders of magnitude more resources.

Electronic nanotechnology is quickly progressing and promises small, dense, low-power devices. Harnessing this power will require new ways of thinking about the manufacturing process. We will no longer be able to manufacture devices deterministically. Instead, post-fabrication reconfiguration will circumvent defects and determine the properties of the device.

Chapter 6

Science and Applications

In this chapter we provide an overview of some key current developments in nanoscience and nanotechnologies, and highlight some possible future applications. The chapter is informed by evidence from scientists and engineers in academia and industry. It illustrates the wide-ranging interest in these areas and provides a background to the later chapters, which address health, environmental, social, ethical and regulatory implications of nanotechnologies.

It does not consider in detail the developments in nanoscience and nanotechnologies in all scientific and engineering fields.

As nanoscience and nanotechnologies cover such a wide range of fields (from chemistry, physics and biology, to medicine, engineering and electronics), we have considered them in four broad categories: nanomaterials; nanometrology; electronics, optoelectronics and information and communication technology; and bio-nanotechnology and nanomedicine. This division helps to distinguish between developments in different fields, but there is naturally some overlap. Where possible, we define the development of future applications as short term (under 5 years), medium term (5–15 years), and long term (over 20years).

It may be that some of the potential applications that we identify are never realised, whereas others that are currently unforeseen could have a major impact. We also identify potential in environmental, health and safety, ethical or societal implications or uncertainties that are discussed further in later chapters.

NANOMATERIALS

Introduction to Nanomaterials

A key driver in the development of new and improved materials, from the steels of the 19th century to the advanced materials of today, has been the ability to control their structure at smaller and smaller scales.

The overall properties of materials as diverse as paints and silicon chips are determined by their structure at the micro- and nanoscales. As our understanding of materials at the nanoscale and our ability to control their structure improves, there will be great potential to create a range of materials with novel characteristics, functions and applications. Although a broad definition, we categorise nanomaterials as those which have structured components with at least one dimension less than 100nm.

Materials that have one dimension in the nanoscale (and are extended in the other two dimensions) are layers, such as a thin films or surface coatings.

Some of the features on computer chips come in this category. Materials that are nanoscale in two dimensions (and extended in one dimension) include nanowires and nanotubes. Materials that are nanoscale in three dimensions are particles, for example precipitates, colloids and quantum dots (tiny particles of semiconductor materials). Nanocrystalline materials, made up of nanometre-sized grains, also fall into this category.

Some of these materials have been available for some time; others are genuinely new. The aim of this chapter is to give an overview of the properties, and the significant foreseeable applications of some key nanomaterials. Two principal factors cause the properties of nanomaterials to differ significantly from other materials: increased relative surface area, and quantum effects. These factors can change or enhance properties such as reactivity, strength and electrical characteristics. As a particle decreases in size, a greater proportion of atoms are found at the surface compared to those inside. For example, a particle of size 30 nm has 5% of its atoms on its surface, at 10 nm 20% of its atoms, and at 3 nm 50% of its atoms. Thus nanoparticles have a much greater surface area per unit mass compared with larger particles. As growth and catalytic chemical reactions occur at surfaces, this means that a given mass of material in nanoparticulate form will be much more reactive than the same mass of material made up of larger particles. In tandem with surface-area effects, quantum effects can begin to dominate the properties of matter as size is reduced to the nanoscale. These can affect the optical, electrical and magnetic behaviour of materials, particularly as the structure or particle size approaches the smaller end of the nanoscale.

Materials that exploit these effects include quantum dots, and quantum well lasers for optoelectronics. For other materials such as crystalline solids, as the size of their structural components decreases, there is much greater interface area within the material; this can greatly affect both mechanical and electrical properties. For example, most metals are made up of small crystalline grains; the boundaries between the grain slow down or arrest the propagation of defects when the material is stressed, thus giving it strength. If these grains can be made very small, or even nanoscale in size, the interface area within the material greatly increases, which enhances its strength.

For example, nanocrystalline nickel is as strong as hardened steel. Understanding surfaces and interfaces is a key challenge for those working on nanomaterials, and one where new imaging and analysis instruments are vital. Nanomaterials are not simply another step in the miniaturization of materials. They often require very different production approaches. There are several processes to create nanomaterials, classified as 'top-down' and 'bottom-up'. Although many nanomaterials are currently at the laboratory stage of manufacture, a few of them are being commercialised.

Nanoscience in this Area

Below we outline some examples of nanomaterials and

the range of nanoscience that is aimed at understanding their properties. As will be seen, the behaviour of some nanomaterials is well understood, whereas others present greater challenges. a) Nanoscale in one dimension

Thin Films, Layers and Surfaces

One-dimensional nanomaterials, such as thin films and engineered surfaces, have been developed and used for decades in fields such as electronic device manufacture, chemistry and engineering. In the silicon integrated-circuit industry, for example, many devices rely on thin films for their operation, and control of film thicknesses approaching the atomic level is routine. Monolayers (layers that are one atom or molecule deep) are also routinely made and used in chemistry.

The formation and properties of these layers are reasonably well understood from the atomic level upwards, even in quite complex layers (such as lubricants). Advances are being made in the control of the composition and smoothness of surfaces, and the growth of films. Engineered surfaces with tailored properties such as large surface area or specific reactivity are used routinely in a range of applications such as in fuel cells and catalysts.

The large surface area provided by nanoparticles, together with their ability to self assemble on a support surface, could be of use in all of these applications. Although they represent incremental developments, surfaces with enhanced properties should find applications throughout the chemicals and energy sectors. The benefits could surpass the obvious economic and resource savings achieved by higher activity and greater selectivity in reactors and separation processes, to enabling small-scale distributed processing (making chemicals as close as possible to the point of use).

There is already a move in the chemical industry towards this. Another use could be the small-scale, on-site production of high value chemicals such as pharmaceuticals. b) Nanoscale in two dimensions Two dimensional nanomaterials such as tubes and wires have generated considerable interest among the scientific community in recent years. In particular, their novel electrical and mechanical properties are the subject of intense research.

Carbon Nanotubes

Carbon nanotubes (CNTs) were first observed by Sumio Iijima in 1991. CNTs are extended tubes of rolled graphene sheets. There are two types of CNT: single-walled (one tube) or multi-walled (several concentric tubes). Both of these are typically a few nanometres in diameter and several micrometres (10-6m) to centimetres long. CNTs have assumed an important role in the context of nanomaterials, because of their novel chemical and physical properties.

They are mechanically very strong (their Young's modulus is over 1 terapascal, making CNTs as stiff as diamond), flexible (about their axis), and can conduct electricity extremely well (the helicity of the graphene sheet determines whether the CNT is a semiconductor or metallic). All of these remarkable properties give CNTs a range of potential applications: for example, in reinforced composites, sensors, nanoelectronics and display devices.

CNTs are now available commercially in limited quantities. They can be grown by several techniques, which are discussed in section. However, the selective and uniform production of CNTs with specific dimensions and physical properties is yet to be achieved. The potential similarity in size and shape between CNTs and asbestos fibres has led to concerns about their safety, which we address in detail in sections..

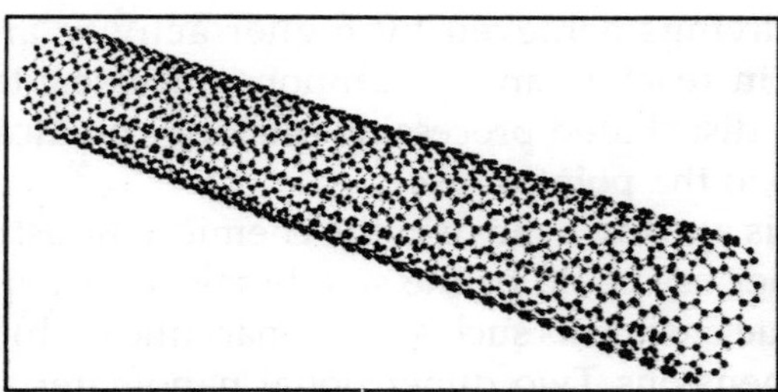

Fig. (a) Schematic of a single-walled Carbon Nanotube (SWNT)

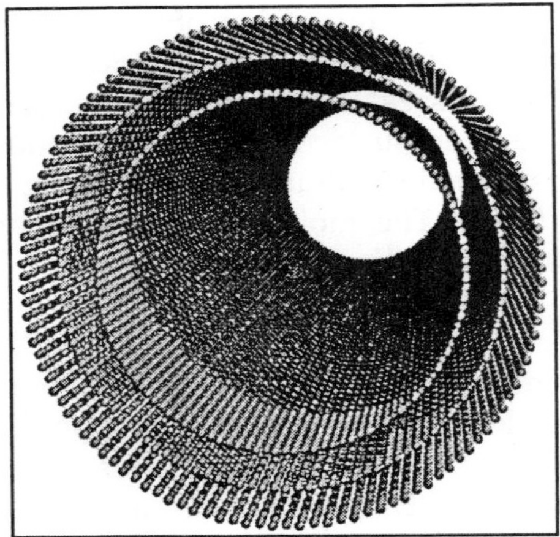

Fig. (b) Schematic of a Multi-Walled Carbon Nanotube (MWNT)

Inorganic Nanotubes

Inorganic nanotubes and inorganic fullerene-like materials based on layered compounds such as molybdenum disulphide were discovered shortly after CNTs. They have excellent tribological (lubricating) properties, resistance to shockwave impact, catalytic reactivity, and high capacity for hydrogen and lithium storage, which suggest a range of promising applications. Oxide-based nanotubes (such as titanium dioxide) are being explored for their applications in catalysis, photo-catalysis and energy storage.

Nanowires

Nanowires are ultrafine wires or linear arrays of dots, formed by self-assembly. They can be made from a wide range of materials. Semiconductor nanowires made of silicon, gallium nitride and indium phosphide have demonstrated remarkable optical, electronic and magnetic characteristics (for example, silica nanowires can bend light around very tight corners). Nanowires have potential applications in high-density data storage, either as magnetic read heads or as patterned storage media, and electronic and opto-electronic nanodevices, for metallic interconnects of quantum devices and nanodevices. The preparation of these nanowires relies on sophisticated growth techniques, which include selfassembly processes, where atoms arrange themselves naturally on stepped surfaces, chemical vapour deposition (CVD) onto patterned substrates, electroplating or molecular beam epitaxy (MBE). The 'molecular beams' are typically from thermally evaporated elemental sources.

Biopolymers

The variability and site recognition of biopolymers, such as DNA molecules, offer a wide range of opportunities for the self-organization of wire nanostructures into much more complex patterns. The DNA backbones may then, for example, be coated in metal. They also offer opportunities to link nano- and biotechnology in, for example, biocompatible sensors and small, simple motors. Such self-assembly of organic backbone nanostructures is often controlled by weak interactions, such as hydrogen bonds, hydrophobic, or van der Waals interactions (generally in aqueous environments) and hence requires quite different synthesis strategies to CNTs, for example.

The combination of one-dimensional nanostructures consisting of biopolymers and inorganic compounds opens up a number of scientific and technological opportunities.

NANOSCALE IN THREE DIMENSIONS

Nanoparticles

Nanoparticles are often defined as particles of less than 100nm in diameter. In line with our definitions of nanoscience and nanotechnologies we classify nanoparticles to be particles less than 100nm in diameter that exhibit new or enhanced size-dependent properties compared with larger particles of the same material.

Nanoparticles exist widely in the natural world: for example as the products of photochemical and volcanic activity, and created by plants and algae. They have also been created for thousands of years as products of combustion and food cooking, and more recently from vehicle exhausts. Deliberately manufactured nanoparticles, such as metal oxides, are by comparison in the minority.

In this report we will refer to these as natural, pollutant and manufactured nanoparticles, respectively. Nanoparticles are of interest because of the new properties (such as chemical reactivity and optical behaviour) that they exhibit compared with larger particles of the same materials. For example, titanium dioxide and zinc oxide become transparent at the nanoscale, however are able to absorb and reflect UV light, and have found application in sunscreens. Nanoparticles have a range of potential applications: in the short-term in new cosmetics, textiles and paints; in the longer term, in methods of targeted drug delivery where they could be to used deliver drugs to a specific site in the body.

Nanoparticles can also be arranged into layers on surfaces, providing a large surface area and hence enhanced activity, relevant to a range of potential applications such as catalysts. Manufactured nanoparticles are typically not products in their own right, but generally serve as raw materials, ingredients or additives in existing products.

Although their production is currently low compared with other nanomaterials we have given them a considerable amount of attention in this report. This is because they are currently in a small number of consumer products such as cosmetics and their enhanced or novel properties may have implications for their toxicity.

The evidence submitted during the course of our study indicates that for most applications, nanoparticles will be fixed (for example, attached to a surface or within in a composite) although in others they will be free or suspended in fluid. Whether they are fixed or free will have a significant affect on their potential health, safety and environmental impacts.

Fullerenes (carbon 60)

In the mid- s a new class of carbon material was discovered called carbon 60 (C60). A diagram of carbon 60 can be found in Figure. These are spherical molecules about 1nm in diameter, comprising 60 carbon atoms arranged as 20 hexagons and 12 pentagons: the configuration of a football.

The C60 species was named 'Buckminsterfullerene' in recognition of the architect Buckminster Fuller, who was well-known for building geodesic domes, and the term fullerenes was then given to any closed carbon cage. In 1990, a technique to produce larger quantities of C60 was developed by resistively heating graphite rods in a helium atmosphere.

Several applications are envisaged for fullerenes, such as miniature 'ball bearings' to lubricate surfaces, drug delivery vehicles and in electronic circuits.

Dendrimers

Dendrimers are spherical polymeric molecules, formed through a nanoscale hierarchical self-assembly process. There are many types of dendrimer; the smallest is several nanometres in size. Dendrimers are used in conventional applications such as coatings and inks, but they also have a range of interesting properties which could lead to useful applications.

For example, dendrimers can act as nanoscale carrier molecules and as such could be used in drug delivery. Environmental clean-up could be assisted by dendrimers as they can trap metal ions, which could then be filtered out of water with ultra-filtration techniques.

Quantum Dots

Nanoparticles of semiconductors (quantum dots) were theorized in the 1970s and initially created in the early 1980s. If semiconductor particles are made small enough, quantum effects come into play, which limit the energies at which electrons and holes (the absence of an electron) can exist in the particles.

As energy is related to wavelength (or colour), this means that the optical properties of the particle can be finely tuned depending on its size. Thus, particles can be made to emit or absorb specific wavelengths (colours) of light, merely by controlling their size.

Recently, quantum dots have found applications in composites, solar cells (Gratzel cells) and fluorescent biological labels (for example to trace a biological molecule) which use both the small particle size and tuneable energy levels. Recent advances in chemistry have resulted in the preparation of monolayer-protected, high-quality, monodispersed, crystalline quantum dots as small as 2nm in diameter, which can be conveniently treated and processed as a typical chemical reagent.

APPLICATIONS

Below we list some key current and potential shortand long-term applications of nanomaterials. Most current applications represent evolutionary developments of existing technologies: for example, the reduction in size of electronics devices.

CURRENT

Sunscreens and Cosmetics

Nanosized titanium dioxide and zinc oxide are currently used in some sunscreens, as they absorb and reflect ultraviolet (UV) rays and yet are transparent to visible light and so are more appealing to the consumer.

Nanosized iron oxide is present in some lipsticks as a pigment but it is our understanding that it is not used by the European cosmetics sector. The use of nanoparticles in cosmetics has raised a number of concerns about consumer safety.

Composites

An important use of nanoparticles and nanotubes is in composites, materials that combine one or more separate components and which are designed to exhibit overall the best properties of each component. This multi-functionality applies not only to mechanical properties, but extends to optical, electrical and magnetic ones. Currently, carbon fibres and bundles of multi-walled CNTs are used in polymers to control or enhance conductivity, with applications such as antistatic packaging.

The use of individual CNTs in composites is a potential long-term application. A particular type of nanocomposite is where nanoparticles act as fillers in a matrix; for example, carbon black used as a filler to reinforce car tyres. However, particles of carbon black can range from tens to hundreds of nanometres in size, so not all carbon black falls within our definition of nanoparticles.

Clays

Clays containing naturally occurring nanoparticles have long been important as construction materials and are undergoing continuous improvement. Clay particle based composites – containing plastics and nano-sized flakes of clay – are also finding applications such as use in car bumpers.

Coatings and Surfaces

Coatings with thickness controlled at the nano- or atomic scale have been in routine production for some time, for example in MBE or metal oxide CVD for optoelectonic devices, or in catalytically active and chemically functionalized surfaces.

Recently developed applications include the self-cleaning window, which is coated in highly activated titanium dioxide, engineered to be highly hydrophobic (water repellent) and antibacterial, and coatings based on nanoparticulate oxides that catalytically destroy chemical agents. Wear and scratch-resistant hard coatings are significantly improved by nanoscale intermediate layers (or multilayers) between the hard outer layer and the substrate material.

The intermediate layers give good bonding and graded matching of elastic and thermal properties, thus improving adhesion.

A range of enhanced textiles, such as breathable, waterproof and stainresistant fabrics, have been enabled by the improved control of porosity at the nanoscale and surface roughness in a variety of polymers and inorganics.

Tougher and Harder Cutting Tools

Cutting tools made of nanocrystalline materials, such as tungsten carbide, tantalum carbide and titanium carbide, are more wear and erosion-resistant, and last longer than their conventional (large-grained) counterparts. They are finding applications in the drills used to bore holes in circuit boards.

SHORT-TERM

Paints

Incorporating nanoparticles in paints could improve their performance, for example by making them lighter and giving them different properties. Thinner paint coatings ('lightweighting'), used for example on aircraft, would reduce their weight, which could be beneficial to the environment. However, the whole life cycle of the aircraft needs to be considered before overall benefits can be claimed. It may also be possible to substantially reduce solvent content of paints, with resulting environmental benefits. New types of foulingresistant marine paint could be developed and are urgently needed as alternatives to tributyl tin (TBT), now that the ecological impacts of TBT have been recognised.

Anti-fouling surface treatment is also valuable in process applications such as heat exchange, where it could lead to energy savings. If they can be produced at sufficiently low cost, fouling-resistant coatings could be used in routine duties such as piping for domestic and industrial water systems. It remains speculation whether very effective anti-fouling coatings could reduce the use of biocides, including chlorine. Other novel, and more long-term, applications for nanoparticles might lie in paints that change colour in response to change in temperature or chemical environment, or paints that have reduced infra-red absorptivity and so reduce heat loss.

Concerns about the health and environmental impacts of nanoparticles may require the need for the durability and abrasion behaviour of nano-engineered paints and coatings to be addressed, so that abrasion products take the form of coarse or microscopic agglomerates rather than individual nanoparticles.

Remediation

The potential of nanoparticles to react with pollutants in soil and groundwater and transform them into harmless compounds is being researched. In one pilot study the large surface area and high surface reactivity of iron nanoparticles were exploited to transform chlorinated hydrocarbons (some of which are believed to be carcinogens) into less harmful end products in groundwater.

It is also hoped that they could be used to transform heavy metals such as lead and mercury from bioavailable forms into insoluble forms. Serious concerns have been raised over the uncontrolled release of nanoparticles into the environment..

Fuel Cells

Engineered surfaces are essential in fuel cells, where the external surface properties and the pore structure affect performance. The hydrogen used as the immediate fuel in fuel cells may be generated from hydrocarbons by catalytic reforming, usually in a reactor module associated directly with the fuel cell.

The potential use of nano-engineered membranes to intensify catalytic processes could enable higher-efficiency, small-scale fuel cells. These could act as distributed sources of electrical power. It may eventually be possible to produce hydrogen locally from sources other than hydrocarbons, which are the feedstocks of current attention.

Displays

The huge market for large area, high brightness, flat-panel displays, as used in television screens and computer monitors, is driving the development of some nanomaterials. Nanocrystalline zinc selenide, zinc sulphide, cadmium sulphide and lead telluride synthesized by sol–gel techniques (a process for making ceramic and glass materials, involving the transition from a liquid 'sol' phase to a solid 'gel' phase) are candidates for the next generation of light-emitting phosphors. CNTs are being investigated for low voltage field-emission displays; their strength, sharpness, conductivity and inertness make them potentially very efficient and long-lasting emitters.

Batteries

With the growth in portable electronic equipment (mobile phones, navigation devices, laptop computers, remote sensors), there is great demand for lightweight, high-energy density batteries.

Nanocrystalline materials synthesized by sol–gel techniques are candidates for separator plates in batteries because of their foam-like (aerogel) structure, which can hold considerably more energy than conventional ones. Nickel–metal hydride batteries made of nanocrystalline nickel and metal hydrides are envisioned to require less frequent recharging and to last longer because of their large grain boundary (surface) area.

Fuel Additives

Research is underway into the addition of nanoparticulate ceria (cerium oxide) to diesel fuel to improve fuel economy by reducing the degradation of fuel consumption over time.

Catalysts

In general, nanoparticles have a high surface area, and hence provide higher catalytic activity. Nanotechnologies are enabling changes in the degree of control in the production of nanoparticles, and the support structure on which they reside. It is possible to synthesise metal nanoparticles in solution in the presence of a surfactant to form highly ordered monodisperse films of the catalyst nanoparticles on a surface. This allows more uniformity in the size and chemical structure of the catalyst, which in turn leads to greater catalytic activity and the production of fewer byproducts.

It may also be possible to engineer specific or selective activity. These more active and durable catalysts could find early application in cleaning up waste streams. This will be particularly beneficial if it reduces the demand for platinum-group metals, whose use in standard catalytic units is starting to emerge as a problem, given the limited availability of these metals. c) Longer-term applications

Carbon Nanotube Composites

CNTs have exceptional mechanical properties, particularly high tensile strength and light weight. An obvious area of application would be in nanotubereinforced composites, with performance beyond current carbon-fibre composites.

One current limit to the introduction of CNTs in composites is the problem of structuring the tangle of nanotubes in a well-ordered manner so that use can be made of their strength. Another challenge is generating strong bonding between CNTs and the matrix, to give good overall composite performance and retention during wear or erosion of composites.

The surfaces of CNTs are smooth and relatively unreactive, and so tend to slip through the matrix when it is stressed. One approach that is being explored to prevent this slippage is the attachment of chemical side-groups to CNTs, effectively to form 'anchors'. Another limiting factor is the cost of production of CNTs. However, the potential benefits of such light, high strength material in numerous applications for transportation are such that significant further research is likely.

Lubricants

Nanospheres of inorganic materials could be used as lubricants, in essence by acting as nanosized 'ball bearings'. The controlled shape is claimed to make them more durable than conventional solid lubricants and wear additives. Whether the increased financial and resource cost of producing them is offset by the longer service life of lubricants and parts remains to be investigated. It is also claimed that these nanoparticles reduce friction between metal surfaces, particularly at high normal loads.

If so, they should find their first applications in high-performance engines and drivers; this could include the energy sector as well as transport. There is a further claim that this type of lubricant is effective even if the metal surfaces are not highly smooth. Again, the benefits of reduced cost and resource input for machining must be compared against production of nanolubricants. In all these applications, the particles would be dispersed in a conventional liquid lubricant; design of the lubricant system must therefore include measures to contain and manage waste.

Magnetic Materials

It has been shown that magnets made of nanocrystalline yttrium–samarium–cobalt grains possess unusual magnetic properties due to their extremely large grain interface area (high coercivity can be obtained because magnetization flips cannot easily propagate past the grain boundaries). This could lead to applications in motors, analytical instruments like magnetic resonance imaging (MRI), used widely in hospitals, and microsensors.

Overall magnetisation, however, is currently limited by the ability to align the grains' direction of magnetisation. Nanoscale-fabricated magnetic materials also have applications in data storage.

Devices such as computer hard disks depend on the ability to magnetize small areas of a spinning disk to record information. If the area required to record one piece of information can be shrunk in the nanoscale (and can be written and read reliably), the storage capacity of the disk can be improved dramatically. In the future, the devices on computer chips which currently operate using flows of electrons could use the magnetic properties of these electrons, called spin, with numerous advantages. Recent advances in novel magnetic materials and their nanofabrication are encouraging in this respect.

Medical Implants

Current medical implants, such as orthopaedic implants and heart valves, are made of titanium and stainless steel alloys, primarily because they are biocompatible. Unfortunately, in some cases these metal alloys may wear out within the lifetime of the patient. Nanocrystalline zirconium oxide (zirconia) is hard, wearresistant, bio-corrosion resistant and bio-compatible. It therefore presents an attractive alternative material for implants.

It and other nanoceramics can also be made as strong, light aerogels by sol–gel techniques. Nanocrystalline silicon carbide is a candidate material for artificial heart valves primarily because of its low weight, high strength and inertness.

Machinable Ceramics

Ceramics are hard, brittle and difficult to machine. However, with a reduction in grain size to the nanoscale, ceramic ductility can be increased. Zirconia, normally a hard, brittle ceramic, has even been rendered superplastic (for example, able to be deformed up to 300% of its original length). Nanocrystalline ceramics, such as silicon nitride and silicon carbide, have been used in such automotive applications as high-strength springs, ball bearings and valve lifters, because they can be easily formed and machined, as well as exhibiting excellent chemical and high-temperature properties. They are also used as components in high-temperature furnaces.

Nanocrystalline ceramics can be pressed into complex net shapes and sintered at significantly lower temperatures than conventional ceramics.

Water Purification

Nano-engineered membranes could potentially lead to more energy-efficient water purification processes, notably in desalination by reverse osmosis. Again, these applications would represent incremental improvements in technologies that are already available. They would use fixed nanoparticles, and are therefore distinct from applications that propose to use free nanoparticles.

Military Battle Suits

Enhanced nanomaterials form the basis of a stateof- the-art 'battle suit' that is being developed by the Institute of Soldier Nanotechnologies at Massachusetts Institute of Technology, USA. A short-term development is likely to be energy-absorbing materials that will withstand blast waves; longer-term are those that incorporate sensors to detect or respond to chemical and biological weapons (for example, responsive nanopores that 'close' upon detection of a biological agent).

There is speculation that developments could include materials which monitor physiology while a soldier is still on the battlefield, and uniforms with potential medical applications, such as splints for broken bones. we consider the possible social implications of the exploitation of nanotechnologies for military purposes.

NANOMETROLOGY

Introduction to Nanometrology

The science of measurement at the nanoscale is called nanometrology. Its application underpins all of nanoscience and nanotechnologies.

The ability to measure and characterise materials (determine their size, shape and physical properties) at the nanoscale is vital if nanomaterials and devices are to be produced to a high degree of accuracy and reliability and the applications of nanotechnologies are to be realised. Nanometrology includes length or size measurements (where dimensions are typically given in nanometres and the measurement uncertainty is often less than 1nm) as well as measurement of force, mass, electrical and other properties.

As techniques for making these measurements advance, so too does the understanding of nanoscale behaviour and therefore the possibility of improving materials, industrial processes and reliability of manufacture. The instruments for making such measurements are many and varied.. As with all measurement, nanometrology is essentially an enabling technology. Nanotechnologies, however defined, cannot progress independently of progress in nanometrology.

Apart from their direct influence on scientific research and its application, the solutions developed for nanometrology problems can often be exploited elsewhere. For example, the concept of the AFM, a key nanometrology tool, has had a direct influence on lithographic processes and techniques for molecular manipulation.

Conversely, it is likely that continuing research into nanodevices will suggest new measurement methods. Making measurements with nanoscale precision poses several major difficulties. Environmental fluctuations such as vibration or temperature change have a large effect at the nanoscale.

For example, any external change to the large machines used in manufacturing microelectronics components will affect the creation of nanoscale features and their crucially important alignment to each other.

The ability to measure these influences, and thereafter to minimise them, is therefore vital. Currently, instruments are available that can make sufficiently precise measurements to support laboratory research. There are a number of sensor technologies and instruments with nanometre, or better, sensitivity for measuring length that repeat well if used carefully, including the scanning probe and electron microscopes and some optical devices.

However, universal measurement standards have not yet been established. Data published recently from the Physikalisch-Technische Bundesanstalt in Germany (Breil et al 2002) shows that even apparently sophisticated users of atomic force microscopes can produce large variations in their measurements of the same artefacts. Without agreed standards, tools or machines cannot be calibrated at the nanometre scale. It is therefore not yet possible for laboratories and manufacturing plants to exchange or compare data or physical components. Also, health and safety standards cannot be set for legal requirements.

Nanoparticle characterization for size, size distribution

and shape is also lacking formal methods. Evidence presented at our industry workshop highlighted that good comparative metrology is proving difficult to develop. There is no particular difficulty with working at the nanoscale within a single laboratory or organization in the sense artifacts (either universal or inhouse 'gold standards' such as the spacing of the silicon lattice) can be used to calibrate instruments so that there is self-consistency across a set of measurements. In doing this there must naturally be levels of protection against vibration, thermal changes, etc that are becoming increasingly stringent.

However, the determination of absolute measurements of length at the nanometer scale and below is very difficult and expensive.

Length Measurement

Although there is not currently an international standard that can be applied to them, calibration artefacts are becoming available for AFMs for length measurement calibration in each of the three dimensions. Such artefacts can also be used for tip characterization, as the exact shape of the fine tip which scans across the surface can strongly influence length measurements, particularly when the tip becomes blunt. The National Physical Laboratory (NPL) in the UK has a considerable reputation in formal metrology and has done much to develop measurement capability as well as to take responsibility for the UK's national standards.

It also has excellent realisations of the metre, mainly through optical and x-ray interferometry, that can calibrate transfer standards with uncertainties of the order of nanometres when working with sufficiently large wave-fronts. However, comparative methods (akin to the interferometric measurement of optical lenses and mirrors) become ineffective when working with very small objects comparable to the wavelength of the photons being used. Hence, it is not yet possible to accurately determine dimensions or shape in all axes.

Many surface characterization methods (especially for the

electronics industries) directly exploit comparison via xray (occasionally optical) reflection or diffraction. It is likely that the practice of length metrology would be improved through better, more easily-used calibration systems and improved instrument design, relative to the current commercial versions. Undoubtedly, better education in best practice for nanometrology would also partly address this issue.

Force Measurement

Along with length measurement, force measurement (measured in Newtons (N)) is likely to become an important area of nanometrology. The control of probe stiffness and geometry will need to improve if truly quantitative measurements of surface mechanical properties can be made, particularly when measuring biological and other soft materials. There is also likely to be an increasing need to accurately measure the elasticity of protein and nucleotide molecules, to determine bond strength and other properties of the molecules. Currently, there is a large capability gap in this field. There is a large, and growing, need for force characterisation in the pico- to micronewton (10^{-12}–10^{-6}N) range.

Currently, no fully satisfactory techniques are available either for secondary standards or transfer artefacts, although a few research projects are in progress (NPL and the National Institute for Standards and Technology (NIST), USA, are both looking at methods based on electrostatic forces). Several groups, mainly within or sponsored by national laboratories (such as NPL and Warwick University in the UK, and NIST in the USA), are investigating systems that relate force to electrical properties and so to quantum standards. However, so far all of them remain experimental and a great deal more work is urgently needed into fundamental and transfer standards for forces much smaller than millinewtons.

Unlike length measurement, there is also a lack of readily available and applicable force or mass instrumentation with sensitivity adequate for engineering on the nanometre scale. AFM cantilevers have nanonewton force sensitivity, but their calibration tends to be through indirect calculation from their dimensions, and batch-tobatch repeatability may be poor. Some nano-indentation instruments for hardness measurement use microelectromechanical systems (MEMS), with broadly similar questions over traceability. Thus there is urgent need for research into basic laboratory and industrial nanoforce instrumentation alongside that for standards.

Measurement of Single Molecules

In the longer term, development in measurement at the scale of the single molecule is expected. Measurements of single organic molecules and of structures such as single-wall nanotubes are already made, providing the molecules can be anchored to a substrate.

Electron microscope and AFM/STM determinations of shape are relatively routine in many research laboratories. Increasingly, there is interest in molecule stiffness, in effect producing a tensile test curve in which jumps indicate the breaking and by inference the location of various types of bond in folded proteins and nucleotides. AFM manufacturers are starting to offer options that can do this without the need for the skills of a large research team.

Applications

Metrology forms the basis of the semiconductor industry and as such is enormously advanced. The International Technology Roadmap for Semiconductors (ITRS) roadmap highlights a series of challenges for nanometrology if it is to keep pace with the reduction in feature size of semiconductor devices. Shrinking feature sizes, tighter control of device electrical parameters and new interconnect materials will provide the main challenges for physical metrology methods. To achieve desired device scaling, metrology tools must be capable of measurement of properties at atomic distances.

Compounding these is the uncertain nature of the development of device design, making it difficult to predict metrology needs in the long term and in particular the necessary metrology for manufacturing to ensure reliability.

A major need is to integrate metrology data into the manufacturing process. The developing capabilities of semiconductor processing, particularly the ever-reducing dimensions that can be defined using lithographic tools, are being combined with the techniques developed for MEMS device fabrication to enable the manufacture of electro-mechanical components with sub-100nm dimensions.

The exploitation of these structures in nano-electromechanical systems (NEMS) devices has produced some interesting and exciting developments in the field of nanometrology. For example, Schwab et al (2000) have made a NEMS device that has enabled the measurement of the quantum of thermal conductance. Another group has made ultra-thin silicon cantilevers with attonewton (10^{-18}N) sensitivity. These devices have potential applications in the characterisation of single molecule properties and are examples of how the field of NEMS is increasing the capabilities of nanometrology. The role of nanometrology and in particular the need for the standardisation of measurement at the nanometre scale is explored further in section 8.4.3.

There is a need to develop agreed standards that can be used to calibrate equipment that will be used by both industry and regulators. We believe that this can best be addressed through existing programmes such as the Department of Trade and Industry (DTI) National Measurement System Programme and should be undertaken in collaboration with industry. We are pleased to learn that initial steps in this area are being undertaken by the British Standards Institution, as part of the European Committee for Standardisation Technical Board working group on nanotechnology.

ELECTRON BEAM TECHNIQUES

Transmission electron microscopy (TEM) is used to investigate the internal structure of micro- and nanostructures. It works by passing electrons through the sample and using magnetic lenses to focus the image of the structure, much like light is transmitted through materials in conventional light microscopes.

Because the wavelength of the electrons is much shorter than that of light, much higher spatial resolution is attainable for TEM images than for a light microscope. TEM can reveal the finest details of internal structure, in some cases individual atoms. The samples used for TEM must be very thin (usually less than 100nm), so that many electrons can be transmitted across the specimen. However, some materials, such as nanotubes, nanocrystalline powders or small clusters, can be directly analysed by deposition on a TEM grid with a carbon support film. TEM and high-resolution transmission electron microscopy (HRTEM) are among the most important tools used to image the internal structure of a sample.

Furthermore, if the HRTEM is adequately equipped, chemical analysis can be performed by exploiting the interactions of the electrons with the atoms in the sample. The scanning electron microscope (SEM) uses many of the basic technology developed for the TEM to provide images of surface features associated with a sample. Here, a beam of electrons is focused to a diameter spot of approximately 1nm in diameter on the surface of the specimen and scanned back and forth across the surface. The surface topography of a specimen is revealed either by the reflected (backscattered) electrons generated or by electrons ejected from the specimen as the incident electrons decelerate secondary electrons.

A visual image, corresponding to the signal produced by the interaction between the beam spot and the specimen at each point along each scan line, is simultaneously built up on the face of a cathode ray tube similar to the way that a television picture is generated. The best spatial resolution currently achieved is of the order of 1nm.

SCANNING PROBE TECHNIQUES

Scanning probe microscopy (SPM) uses the interaction between a sharp tip and a surface to obtain an image. The sharp tip is held very close to the surface to be examined and is scanned back-and-forth. The scanning tunnelling microscope (STM) was invented in 1981 by Gerd Binnig and Heinrich Rohrer, who went on to collect the Nobel Prize for Physics in

1986. Here, a sharp conducting tip is held sufficiently close to a surface (typically about 0.5nm) that electrons can 'tunnel' across the gap. The method provides surface structural and electronic information with atomic resolution. The invention of the STM led directly to the development of other 'scanning probe' microscopes, such as the atomic force microscope. The atomic force microscope (AFM) uses a sharp tip on the end of a flexible beam or cantilever. As the tip is scanned across the sample, the displacement of the end of the cantilever is measured, usually a laser beam.

Unlike the STM, where the sample has to be conductive, an AFM can image insulating materials simply because the signal corresponds to the force between the tip and sample, which reflects the topography being scanned across. There are several different modes for AFM. In contact mode, the tip touches the sample; this is simple to implement but can lead to sample damage from the dragging tip on soft materials. Tapping mode mitigates this difficulty: the tip is oscillated and only touches intermittently, so that dragging during scanning is minimized. Non-contact mode is where the tip senses only the attractive forces with the surface, and causes no damage.

It is technically more difficult to implement since these forces are weak compared with contact forces. In non-contact mode at larger tip-surface separation, the imaging resolution is poor, and the technique not often used. However, at small separation, which requires specialized AFM apparatus to maintain, true atomic resolution can be achieved in non-contact mode AFM.

OPTICAL TWEEZERS (SINGLE BEAM GRADIENT TRAP)

Optical tweezers use a single laser beam (focused by a high-quality microscope objective) to a spot on a specimen plane. The radiation pressure and gradient forces from the spot creates an 'optical trap' which is able to hold a particle at its centre. Small interatomic forces and displacements can then be measured. Samples that can analysed range from single atoms and micrometre-sized spheres to strands of DNA and living cells. Optical tweezers are now a standard method of manipulation and measurement. Numerous traps can be used simultaneously with other optical techniques, such as laser scalpels, which can cut the particle being studied.

ELECTRONICS, OPTOELECTRONICS AND INFORMATION AND COMMUNICATION TECHNOLOGY (ICT)

Introduction to Electronics, Optoelectronics and ICT

The past years has seen a revolution in information technology (IT) that has impacted the lives of many people around the world. At the heart of this revolution is the desire to share information, whether the printed word, images or sounds. This requires a technology that can absorb and process information on one side of the planet and deliver it almost instantaneously to the other in a form that is immediately accessible. Such a technology places enormous pressure on advances in processing and storing information, and on transmitting it and converting it from and to a human readable form.

It also increasingly requires secure encryption of information so that access to information can ultimately be restricted to particular individuals. The market size of the IT industry is currently around $1000 billion, the order of $150 for every human being on the planet, with an expectation that it will reach $3000 billion in 2020. In no other industry sector is the trend for miniaturisation so apparent. This is perhaps most obvious by charting the number of transistors, the building blocks of computer chips, over the past 30years. In 1971 there were just 2300 transistors on Intel's 4004, their first computer chip, with a clock speed (a measure of how fast the chip could operate) of 0.8 million cycles per second.

By 2003 the Intel Xeon processor had 108 million transistors operating at clock speeds in excess of 3,000 million cycles per second. Remarkably, the physical size of the computer chip has remained virtually unchanged over this time; it is the transistor and all the circuitry associated with it that has shrunk dramatically. The increase in the number of transistors on a chip coupled with increased speed have fuelled the economics of the IT industry; in 1971 the fabrication of a single transistor cost about 10 cents; it is currently less than one-thousandth of a cent.

This evolutionary progression of technology is charted and anticipated in the ITRS roadmap, a worldwide consensus-based document that predicts the main trends in the semiconductor industry 15 years into the future. The roadmap which defines in detail all elements of technology that have to be realised for each step change improvement in manufacturing process. This roadmap is used by all industries that are directly or indirectly involved in the manufacture of silicon chips. It identifies material, architecture, metrology and process challenges as well as addressing environmental and heath issues in manufacturing.

Nanoscience in this Area

Nanoscience research in ICT shares many of the same goals as for other applications of nanotechnologies: an improved understanding of nanoscale properties of materials and devices, advances in fabrication and process technology to satisfy increasingly stringent dimensional tolerances, and exploration of alternative technologies that may offer economic or performance benefit.

There is no doubt that the ICT sector has effectively driven a large proportion of nanoscience. Indeed, the first use of the word nanotechnology was in relation to ultra thin layers of relevance to the then upand- coming semiconductor industry. Since then, the research into all aspects of semiconductor device fabrication, from fundamental physics to process technology, has dominated the nanoscience landscape and will continue so to do.

Decreasing device scales will add further impetus to the truly nanoscalar aspects of this global research activity. The ICT sector is, and for historical and economic reasons is likely to remain, heavily silicon-based for the foreseeable future. However, the end of the ITRS roadmap, currently set at 2018 (commonly referred to as the end of Moore's Law) has prompted intensive research into alternative or hybrid technologies for electronics such as conducting polymers, which are discussed further below.

CURRENT APPLICATIONS

Computer Chips

The dominant role of miniaturisation in the evolution of the computer chip is reflected in the fact that the ITRS roadmap defines a manufacturing process standard – a technology node – in terms of a length. The current 130 nm technology node that produces the Intel Xeon processor defines the size of the DRAM (dynamic random access memory) half-pitch (half the distance between two adjacent metal wires in a memory cell). This is turn places a requirement on the lithography, process technology and metrology required to manufacture a working device to this tolerance.

As a comparison, the 1971 Intel 4004 chip used 10,000 nm technology; the chips of 2007 and 2013 will require 65nm and 32nm technology, respectively. In the broadest sense, computer chips in current manufacture are therefore already using nanotechnologies and have been so doing for over 20 years. Furthermore, it is not simply the DRAM half-pitch that is on the nanometre scale.

All the technology that goes into the research, metrology and production of chips has been working, in some cases, at the sub-nanometre atomic level. The variety of tools that support the IT industry includes computer modelling of advanced devices and materials atom by atom, microscopies that can image single atoms, metrologies that can define the absolute position of a single atomic defect over a 30cm diameter wafer (the substrate used for computer chips), thin-film growth processes that can produce layers of material with atomic precision, and lithographies that can 'write' features, such as the DRAM cell, with an accuracy of sub-10nm.

Information Storage

A technology that has necessarily developed in tandem

with IT is that of memory for data storage. This can be divided into two quite different types: solid-state memory such as DRAM that a processor chip would use or flash memory for storing images in a digital camera; and disk-based memory such as the magnetic hard drives as found in all computers. Solid-state memory essentially uses the same processes and technology as the computer chip, with very similar design rules and a similar emphasis on packing more memory into a given area to increase total memory per device.

The development of the hard disk drive, however, has taken a quite different route in evolution as it is based on reading and writing information magnetically to a spinning disk. It is therefore primarily mechanical, or more strictly electro-mechanical, and presents quite different technical challenges. Once again, however, the importance of length scales is paramount as the ideal disk drive is one that has the minimal physical size with a massive ability to store data. This is reflected in the evolution of the disk drive over the past 50 years. The first magnetic hard drive was developed by IBM in 1956 and required fifty 24 inch disks to store five megabytes (million bytes) of data.

In 1999 IBM introduced a 73gigabyte (thousand million bytes) drive that could fit inside a personal computer; that is, over 14,000 times the available data storage in a device less than onethousandth the size of the 1956 drive. Although the individual bits of magnetic information that are written onto the disk drive to give it the high-density storage are currently smaller than 100 nm, the constraints related to this nanotechnology on other aspects of the drive require fabrication of components with even greater precision. The importance of this nanotechnology in the related compact disk (CD) and digital versatile disk (DVD) drives that are now commonplace is equally ubiquitous.

Optoelectronics

The other crucial element of the IT revolution, optoelectronics, relates to devices that rely on converting electrical signals to and from light for data transmission, for displays for optical-based sensing and, in the future, for optical-based computing. Technology in this sector is strongly associated with those described above, and relies substantially on the tools developed there. Although some optoelectronic devices do not depend so critically on miniaturisation as computer chips do, there is nevertheless a similar trend towards miniaturisation, with some existing components, such as quantum-well lasers and liquid crystal displays, requiring nanometre precision in their fabrication.

APPLICATIONS ANTICIPATED IN THE FUTURE

The future development of hardware for the IT industry can be conveniently separated into two paths: a path that is following the well-established ITRS roadmap (which projects out to 2018); and a path that explores alternative technologies and materials that may supersede the roadmap. For the roadmap, miniaturisation remains a key driving force, so that a 22nm technology node is envisaged for manufacture in 2016. Having set this technology target, it is possible to anticipate all the challenges associated with realising it.

Such challenges are detailed extensively in the roadmap but include enhancing performance by introducing new materials such as low dielectrics and higher-conductivity interconnects (wiring), developing lithographies capable of fabricating structures in the sub-50nm range, and integrating advanced metrology tools into the manufacturing process capable of detecting and sizing defects down to the nanometre size. As such, nanoscience and nanotechnologies will continue to have a pivotal role in developing new generations of chips. Related technology such as flash memory will evolve in a similar fashion, with the aim of maximising memory capacity in the smallest possible device. Hard disk technologies, although not explicitly part of the ITRS roadmap, will continue to increase in memory density.

However, there are prospects for some step changes in technology that may significantly change the data storage industry. One obvious potential trend is for solid-state memory to replace disk-based memory. This is already obvious in, for example, personal music players where, as solid-state memory increases in density, hard-disk-based storage is competing with the jog-proof solid-state players. It is likely that the hard disk, whether magnetic or optical, will still be the choice for large volume data storage for the foreseeable future, especially as the bit size shrinks even further.

This is an active area of nanoscience research. Optoelectronics, although not as dependent upon length-scale tolerances as computer chips and data storage, will nevertheless have challenges of its own. Integration of optical components into silicon devices has started and can be expected to evolve further. Some of the challenges where nanotechnologies will have an impact will be in the area of photonic band-gap materials, where the propagation of light through a device can be controlled with the aim of computing with light.

Photonic crystals, fabricated either through a lithographic process or through a self-assembly technique, confine light into precisely controlled pathways in a device structure so that both transmission and functionality can be combined into a single structure. A typical photonic crystal would consist of an array of holes in a dielectric material, fabricated with sub-10nm accuracy, so that the periodicity of the holes determines the ability of the material to transmit the light at any given wavelength. The development of photonic crystals could mean that optical integrated circuits are shrunk further, making a significant impact in areas such as communications and optical computing. Quantum computing and quantum cryptography will also benefit from advances in optoelectronics.

Both technologies rely on the fact that discrete energy (quantum) levels increasingly dominate as electromagnetic energy is confined into smaller and smaller structures. Assuming that the considerable technological challenges of making nanostructures from complex materials can be solved, in some cases by designing at the level of the single atom, then on a 10year time-scale quantum cryptography (a much more secure encryption technology) will replace current encryption methods.

On a similar time-scale, quantum computing will start to provide solutions to complex problems that are difficult or impossible to solve by conventional computing. Once controlling where, how and when light interacts is possible by the advances in technology alluded to above, there is the potential for developing new types of optical spectroscopy at the level of the single molecule, assembling nanostructures by arrays of optical tweezers placing objects into patterns on surfaces, new optical lithographic methods for fabricating computer chips, and optical devices that act as biosensors with detection of single molecules.

The last type of sensor, able to detect the presence of a single molecule in, say, a drop of blood, represents one of the greatest challenges for nanotechnologies. Not only does it require precision in manufacture, but it also requires a unique mixture of electronics, optics, chemistry, biochemistry and medicine to make devices that can be used routinely, cheaply and reliably to monitor the state of human health.

An example of this is in point-of-care health screening where a single drop of blood placed on a sensor chip would be almost instantaneously analysed to provide data to aid a diagnosis. This will require the processing power of a silicon chip with biochemical sensitivity to identify many blood components. This type of monitoring could also begin to be incorporated within the body to provide constant monitoring of health, such as in the control of diabetes or in critical care.

There are many other potential applications of such devices in medicine, making this an area of increasing investment. Alternative, 'off-roadmap' technology will have a similar reliance on nanoscience and nanotechnologies as that of the IT sector described above, but with far greater freedom to explore materials and architectures that may have little resemblance to existing technology. Plasticbased electronics is an example of an alternative technology. It does not directly compete with siliconbased devices but, because of its vastly cheaper fabrication, offers a far cheaper alternative. For inexpensive electronic and optoelectronic applications where speed and high memory density are less important, such as smart cards, plastic offers a new approach to building electronics. Plastic-based electronics is already moving into the commercial sector with potential for considerable growth. Similarly, the use of single molecules as functional elements in future circuits will continue to be an important element of nanoscience where the size of the molecule, typically less than 1nm, offers the ultimate in miniaturisation. In fact, the goal of shrinking function down to single molecules and atoms, foreseen by Richard Feynman in 1959, is the only way to go beyond the currently foreseen evolutionary limit of the ITRS roadmap; conventional silicon transistors have a size limit of the order of tens of nanometres. Nanoscience is still pursuing the concept of storing and processing information at the atomic scale with the hope of, for example, quantum computing and atomic memory where each bit of data is stored on a single atom.

Sensors

Nanotechnologies play several important roles in developing sensor technology. First, the ideal sensor will be minimally invasive and therefore as small as possible. This includes the power supply, the sensing action, whereby the detected property is converted into an electrical signal, and the transmission of the sensing signal to a remote detector. Combining these actions into a device that is smaller than 1mm2 will certainly require nanofabrication techniques, similar to those employed by the IT industry. The second role for nanotechnologies will be in designing the sensing element to be as specific and accurate as possible; as the sensor dimension decreases the area of the sensor available to effect detection will also decrease, making increasing demands on sensitivity.

In the limit of, say, chemical detection this may require detection at the single molecule level; close to the bottom end of the nanotechnology length scale and a significant technical challenge. Nanotechnologies are therefore expected to enable the production of smaller, cheaper sensors with increasing selectivity, which can be used in a wide range of applications.

These include monitoring the quality of drinking water, measuring mechanical stresses in buildings or vehicles to monitor for structural damage, detecting and tracking pollutants in the environment, checking food for edibility, or continuously monitoring health.

Developments could also be used to achieve greater safety, security, and individualised healthcare, and could offer advantages to business (for example in tracking and other monitoring of materials and products). However, there are concerns that the same devices that are used to deliver these benefits might also be used in ways that limit privacy of groups or individuals. Other potential applications vary from monitoring the state and performance of products and materials to give early warning of the need for repair or replacement to enhancing human capabilities by extending physical performance.

BIO-NANOTECHNOLOGY AND NANOMEDICINE

Introduction to Bio-Nanotechnology and Nanomedicine

Without doubt the most complex and highly functional nanoscale machines we know are the naturally occurring molecular assemblies that regulate and control biological systems. Proteins, for example, are molecular structures that possess highly specific functions and participate in virtually all biological sensory, metabolic, information and molecular transport processes. The volume of a single molecule bionanodevice such as a protein is between one-millionth and one-billionth of the volume of an individual cell. In this respect the biological world contains many of the nanoscale devices and machines that nanotechnologists might wish to emulate. Bio-nanotechnology is concerned with molecularscale properties and applications of biological nanostructures and as such it sits at the interface between the chemical, biological and the physical sciences.

It does not concern the large-scale production of biological material such as proteins or the specific genetic modification of plants, organisms or animals to give enhanced properties.

By using nanofabrication techniques and processes of molecular self-assembly, bio-nanotechnology allows the production of materials and devices including tissue and cellular engineering scaffolds, molecular motors, and biomolecules for sensor, drug delivery and mechanical applications. Bionanotechnology can be used in medicine to provide a systematic, as well as a screening, approach to drug discovery, to enhance both diagnostic and therapeutic techniques and to image at the cellular and sub-cellular levels, at a much higher resolution than that of magnetic resonance imaging (MRI).

Nanoscience in this Area

The primary aim of much current research is to obtain a detailed understanding of basic biochemical and biophysical mechanisms at the level of individual molecules. This knowledge will allow the design rules of naturally occurring molecular machines to be determined, which may lead to new technological applications. As we saw in section 3.3, several tools have developed in recent years, such as SPM, that allow the direct observation of the behaviour of single molecules within biological systems. Examples range from the relatively large (45nm) rotary molecular motors that power bacterial flagella 'propellers' to the tiny enzymes such as ATP-synthase (9nm) that catalyse energy conversion in biological processes.

The intricate sequence of changes in molecular structure that forms the basis of such biomolecular machines can now be measured directly by using AFM and 'optical tweezers'. The recent development of highspeed AFM has enabled real-time molecular movement within a molecular motor to be observed directly.

Future bio-nanotechnology and nanomedicine devices may exploit many classes of functional biological materials. One particular group of proteins that is attracting attention are the membrane proteins; these are another class of protein-based machine that regulate many physiological processes. They include ion channels that enable rapid yet selective flux of ions across the cell membrane, hormone receptors that behave as molecular triggers, and photoreceptors that switch between different conformational states by the absorption of a single photon of light, the process that is the basis of vision and photosynthesis.

That approximately onequarter of all genes code for membrane proteins provides evidence of their immense biological importance; it is estimated that they will be the target of up to 80% of all new drugs. Single molecule techniques for both observation and manipulation are now being used routinely to study the selectivity and gating mechanisms of ion channels, and their response to drugs.

CURRENT AND FUTURE APPLICATIONS

Bio-nanotechnology is regarded by many experts as a longer-term prospect: much fundamental science must first be investigated, and many applications, especially in the medical field, will by necessity have to undergo strict testing and validation procedures. The time-scale for such applications is 10 years and beyond.

In the shorter term it may be possible to use proteins, DNA and other bio-polymers directly in nanoelectronics and biosensor applications, but factors such as biocompatibility and robustness may prove to be serious obstacles. Alternatively, bio-mimetic structures may be devised that are based on naturally occurring machines: examples include catenanes and rotaxanes, compounds that behave as rotary or linear molecular motors, respectively. Applications in the field of medicine are especially promising. Areas such as disease diagnosis, drug delivery and molecular imaging are being intensively researched.

Medical-related products containing nanoparticles are currently on the market in the USA. Examples that exploit the known antimicrobial properties of silver include wound dressings containing nanocrystalline silver, which release ionic silver over a sustained period of time to provide a claimed extensive antimicrobial spectrum of 150 different pathogens. a) Array technologies The enormously powerful array technologies, which use relatively large biological samples at the micrometre scale, are continuously being enhanced for sensitivity, size and data analysis. The original DNA chip approach, which carries an array of DNA molecules on an inert carrier, is now routinely used in gene and protein analysis. The push towards higher resolution and smaller sample volume makes this an emerging nanotechnology.

Lab-on-a-chip technologies, which are used for sensing and supporting disease diagnosis, are also currently in the micrometre range, but progress in nanofluidic systems will potentially lead to integrated nanoscale systems becoming available. These could have a range of applications, for example in improved devices for detection of biological and chemical agents in the field.

Electronics and Information and Communication Technology

One of the objectives of bio-nanotechnology research is to use the highly specialised functionality of proteins in devices such as molecular sensors. One of the greatest challenges is to understand the fundamental electronic properties of such molecules and the mechanisms by which electronic charge is transferred between them and metals, semiconductors and novel nanoelectronic components such as CNTs. Progress in this area could allow these 'smart' molecules to be integrated into devices and networks for specific or indeed ICT applications: the realisation of a proteinbased transistor is a major scientific challenge.

DNA itself may turn out to be a useful electronic material, although the weight of experimental evidence indicates that it is not a good electrical conductor; however, used as a template, gold or silver 'coated' DNA nanowires can be produced, and integrated circuits using DNA interconnects have already been realised which use the information coded in the DNA.

Thin films and crystals of the membrane protein bacteriorhodopsin have already been demonstrated to have potential photonics applications such as optically addressable spatial light modulators, holographic memories and sensors.

The photosynthetic reaction centre in this protein, which is only 5 nm in size, behaves as a nanometre diode and so it may be useful in single molecule optoelectronic devices. For example, its integration with electrically conducting CNTs and nanometre electrodes could lead to logic devices, transducers, photovoltaic cells, memories and sensors.

Self-assembly

The top-down approach to nanofabrication has the advantage that almost any pre-determined structure can be produced. However, much attention is now being focused on processes that involve some degree of molecular self-assembly, and in this respect biological materials have remarkable advantages over inorganic materials in the diversity of self-assembled structures that they can produce. Evolution in the natural world has produced an astonishing variety of biomolecular devices, and compared with conventional technologies, many natural molecular devices display enormous functionality.

Among the most outstanding examples of synthetic structures now being fabricated are DNAbased geometrical structures (including artificial crystals) and functioning DNA-based nanomachines.

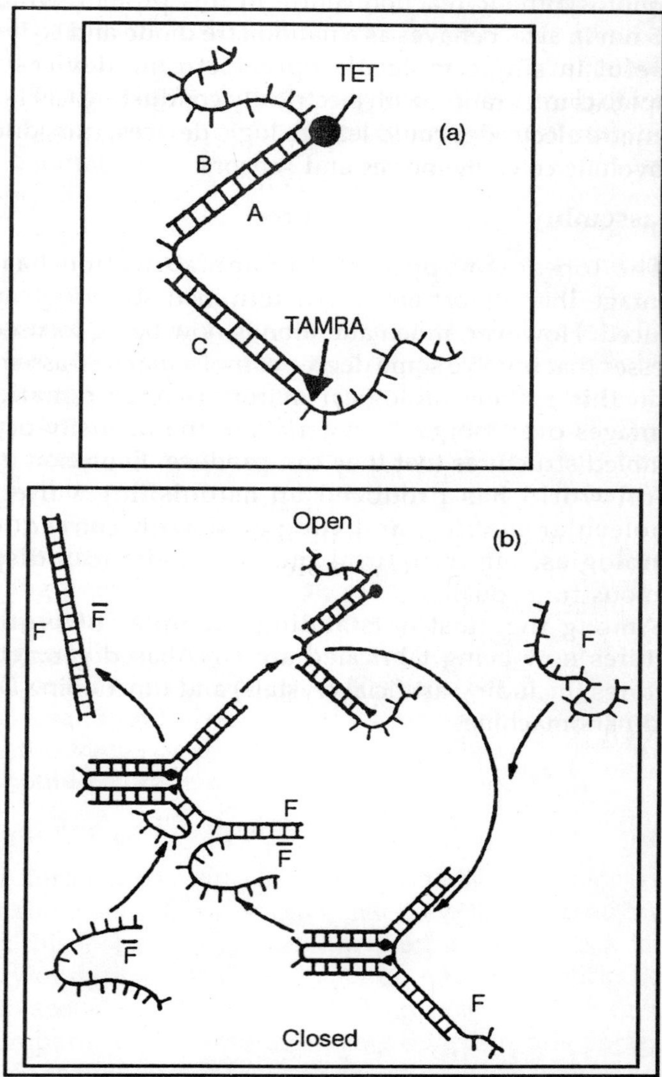

Fig. DNA Nanomachine

DNA Nanomachine (a) A simple device composed of three short single strands of DNA can be made to operate as a tweezer that opens and closes on the addition of another strand. The base sequences are chosen to make parts of A and

B and parts of A and C complementary with each other so that double strands form; this produces the tweezer that is initially in the open state. (b) The addition of a strand F that is complementary to the unpaired sections of B and C causes the tweezer to close when pairing occurs. The tweezer opens again when a strand Fbar is added that is complementary to F: Fbar pairs with F to form a doubled stranded DNA by-product. The energy source for the machine is the hybridisation energy of the FFbar by-product.

Hybrid nanomachines, composed of biological material with inorganic components, have been suggested as posing a threat if they are able to replicate. There are ongoing investigations into the application of biological machines that involve incorporation or transport of non-biological components or material, but these are basic molecular constructions compared with even a simple cell. Although they have the ability to move when chemical fuel is added, the working group found no convincing evidence that self-replication, a characteristic of a living organism, is possible.

Drug Delivery

There is enormous potential for nanotechnology to be applied to gene and drug delivery. The vehicle might be a functionalised nanoparticle capable of targeting specific diseased cells, which contains both therapeutic agents that are released into the cell and an on-board sensor that regulates the release.

Different stages of this approach have already been demonstrated, but the combined targeting and controlled release have yet to be accomplished. In this event the way will be opened up for initial trials, and the eventual approval of such techniques will be fully regulated as for any new pharmaceutical. A related approach already in use is that of polymerbased drug therapies: they include polymeric drugs, polymer–drug conjugates, polymer–protein conjugates, polymeric micelles to which the drug is covalently bound, and multi-component complexes being developed as non-viral vectors for gene therapy.

Many of these materials are now undergoing clinical trials for a variety of disease states.

Gene therapy, where the DNA has been packaged into a nanometre-scale particle, holds much promise for the treatment of genetic defects such as cystic fibrosis and immune deficiencies. Gene therapy using viral vectors has been successfully used to treat ten children with severe combined immunodeficiencies. These are life threatening diseases for which there is no alternative treatment.

Unfortunately two of the children subsequently developed leukaemia leading to a temporary moratorium on all gene therapy trials in 2003. After intensive risk assessments most trials have now resumed.

Alternative non-viral approaches bionanotechnology approaches are being actively researched although none has reached clinical trials.

Advantages of these approaches include the versatility of synthetic chemistry, which allows tailoring of molecular weight, addition of biomimetic features to the manmade construct and even the possibility to include bioresponsive elements.

Drug Discovery

Nanotechnology techniques offer the possibility of studying drug–receptor interactions at the single molecule level, for example by using optical tweezers and AFM, so that a more direct approach to drug discovery becomes feasible. This approach might also allow, for example, the discovery of disease at the single cell level, long before physical symptoms are manifested. This has been achieved by monitoring changes in atomic forces or ion conductance of a single receptor or ion channel when a drug molecule attaches.

However, the industrial process will require the development of large arrays of such instruments working in parallel to create a high-throughput screening capability.

Medical Imaging

Non-invasive imaging techniques have had a major

impact in medicine over the past 25 years or so. The current drive in developing techniques such as functional MRI is to enhance spatial resolution and contrast agents. Nanotechnologies already afford the possibility of intracellular imaging through attachment of quantum dots or synthetic chromophores to selected molecules, for example proteins, or by the incorporation of naturally occurring fluorescent proteins which, with optical techniques such as confocal microscopy and correlation imaging, allow intracellular biochemical processes to be investigated directly.

Nanotechnologies and Cancer Treatment

In the USA the National Nanotechnology Initiative has claimed that nanotechnology has potential in the treatment of cancer. It has been stated that 'It is conceivable that by 2015, our ability to detect and treat tumors in their first year of occurrence might totally eliminate suffering and death from cancer'. We have, however, seen no evidence to support the notion that nanotechnologies will eliminate cancer in the short- to medium term, and feel that such a claim demonstrates an over-simplistic view of the detection and treatment of cancer.

Although it is reasonable to hope that some measures based on nanotechnologies may make contributions to detection and treatment of some forms of cancer, other factors such as a greater understanding of environmental causes of cancer, public health measures, and advances in surgical, pharmacological and radiological management are important in the reduction of incidence of and death from cancer.

Implants and Prosthetics

As discussed in section., some nanomaterials such as nanocrystalline ceramics have certain properties – such as hardness, wear resistance and biocompatibility – that may make them of use as implants in the long term. The development of nanoelectronic systems with high detector densities and data processing capability might allow the development of an artificial retina or cochlea. Important progress is already being made in this area, but many issues must be resolved before they can become viable treatments. Similarly, the introduction of nanoelectronics will allow biological neural processing to be investigated at much enhanced spatial resolution.

Neurons of rodents have already been grown on nanofabricated surfaces to form elementary neural networks in which electrical signalling can be measured. By sending and receiving electrical impulses from the network, it might begin to be possible to understand how neurons create memory by their responses to different patterns of stimuli. It is hoped that this research might help some visually impaired people regain their sight, or that muscle function might be restored to sufferers of Parkinson's disease.

However, these developments raise potential ethical concerns about human enhancement and the convergence of technologies, in particular whether the availability of body alterations that enhance human performance might diminish the role of disabled people in society, and whether progress in information processing and data storage technologies combined with developments in neurophysiology could lead to the possibility of non-therapeutic enhancement of human performance.

Chapter 7

Towards Nanocomputer Architecture

At the nanometer scale, the focus of micro-architecture will move from processing to communication. Most general computer architectures to date have been based on a "stored programme" paradigm that differentiates between memory and processing and relies on communication over busses and other (relatively) long distance mechanisms. Nanometer-scale electronics – nanoelectronics - promises to fundamentally change the ground-rules. Processing will be cheap and plentiful, interconnection expensive but pervasive. This will tend to move computer architecture in the direction of locallyconnected, reconfigurable hardware meshes that merge processing and memory.

If the overheads associated with reconfigurability can be reduced or even eliminated, architectures based on non-volatile, reconfigurable, finegrained meshes with rich, local interconnect offer a better match to the expected characteristics of future nanoelectronic devices. Computer designers have traditionally had to trade the performance of a machine for the area occupied by its component switches. However, when the first practical "nano" scale devices - those with dimensions between one and ten nanometers (10 to 100 atomic diameters) - start to emerge from research laboratories within two or three years, they will mandate a new approach to computer design.

Montemerlo et al (1996) have described the greatest challenge in nanoelectronics as the development of logic designs and computer architectures necessary to link small, sensitive devices together to perform useful calculations efficiently. Ultimately, the objective is to construct a useful "Avogadro computer" - one with an architecture that makes efficient use of in the order of 1023 switches to perform computations. In the more immediate term, it is forecast that by 2012 a CMOS (or possibly SiGe) chip will comprise almost 1010 transistors and will operate at speeds in the order of 10 - 15GHz.

The design challenges will be formidable. For example, amongst a long list of major technical difficulties, the SIA roadmap (which refers particularly to CMOS) identifies the following major issues:

- Power management at all levels;
- New architectures to overcome bottlenecks at interconnects;
- Ultimate short channel limitations (e.g. at 30nm) requiring more complex gate structures such as SOI or dual-gate transistors;
- The spiralling costs of both lithography and fabrication.

In addition to the fundamental problems caused by high power density physical problems such as leakage, threshold voltage control, tunnelling, electromigration, high interconnect resistance, crosstalk and the need for robust and flexible error management become significant as device features shrink.

These problems, in turn, affect the way that devices may be connected together and will ensure that the performance of future architectures will come to be dominated by interconnection constraints rather than by the performance of the logic.

It is likely, therefore, that the physics of nanoelectronic devices will conspire to eliminate the classical storedprogram architecture as a contender at nanoelectronic device densities. This organisation, which has driven the development of computer architecture for almost 50 years, differentiates between the functions of memory and logic processing and tends to be built around system-wide constructs such as busses and global control signals.

It is hard to imagine how any form of globally connected stored-programme architecture could be built in a technology where communication even between adjacent switches is difficult.

Nevertheless, if the progress implied by Moore's Law is to continue nanocomputer architectures must eventually supersede conventional, general-purpose microprocessor machines. They will therefore need to perform the same functions as their predecessors as well as sharing many of their overall characteristics.

They will (ideally) need to be small, fast, cheap and robust, work at room temperature and run code from a standard compiler, including legacy code. This legacy requirement is often overlooked. It is likely that computing functions will continue to be described in terms of software with its inherently linear control flow. General purpose computing is dominated by control dependencies and tends to rely on dynamic data structures.

How the temporal "controlflow" and dynamic data allocation of such a software description might be mapped efficiently onto the hardware circuits of a nanocomputer is not yet clear.

SCALING LIMITS OF CMOS

CMOS has been the work-horse technology in commercial VLSI systems for about 10 years, after superseding nMOS in the early 1990's. During that time, transistor channel lengths have shrunk from microns down to today's typical dimensions of 150 to 180nm and are certain to further scale to 70- 100nm in the near future.

Such devices have already built on research lines – for example by Asai and Wada (1997), Taur et al (1997) and Tang et al (2001) - and these experiments have demonstrated that massproduction is possible. In order to contain an escalating power-density and at the same time maintain adequate reliability margins, traditional CMOS scaling has relied on the simultaneous reduction of device dimensions, isolation, interconnect dimensions, and supply voltages.

However, FET scaling will be ultimately limited by high fields in the gate oxide and the channels, short channel effects that reduce device thresholds and increased subthreshold leakage currents.

As a result, Davari has suggested that gains in FET device performance will eventually stall as the minimum effective channel length approaches 30nm at a supply voltage of 1.0V and a gate oxide thickness of about 1.5nm. Beyond this point, any further performance growth will need to rely on increased functional integration with an emphasis on circuit and architectural innovations.

Defect and Reliability Limits

The probability of failure for transistors in current CMOS manufacturing processes range from 10-9 to 10-7 and it appears certain that currently available processes will not be suitable for providing defect-free device structures at sub-100nm scales.

Thus any architecture built from large numbers of nanoscale components will necessarily contain a significant number of defects.

An understanding of the role of these defects and how they affect yield will be important to future architectures. Novel low-temperature, 3-D integrated manufacturing technologies such as that proposed by Ohmi et al (2001) might eventually result in reliable, defect-free, high-performance gigahertz-rate systems.

However, given the investment in current silicon processing lines, there is no reason to expect that these will be available soon, or that defect rates on typical process lines will improve more than an order of magnitude moving into the nanometre region.

Thus, defects are guaranteed to remain a major technical issue at the architectural level. A closely related problem is the longer term reliability of nanoelectronic technology.

The reliability curve developed for ULSI logic by Shibayama et al (1997) indicates that at gate densities in the order of 107 almost half of systems can be expected to have failed within 10 years (based on the assumption that a single gate failure results in the failure of the entire system). Extrapolating these curves for transistor densities in the order of 109 (the IST forecast for 2006) would imply a 90% failure rate within about 1.3 years1.

To maintain the same reliability as a 1 million gate chip would require an error rate in the order of 10-16/hour-gate, four orders of magnitude better than current technology. How these curves might eventually be extended to a system with 1023 devices is unclear.

What is clear, however, is that nanocomputer architectures will certainly need to be dynamically defect tolerant - with an ability to find defects as they develop and to reconfigure around them (Koren and Koren 1998).

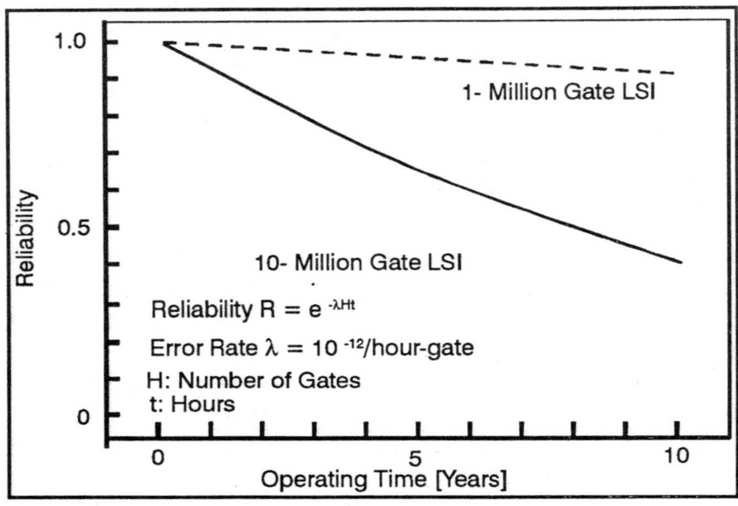

Fig. ULSI Reliability Curves

As a result, testing will represent a major issue in nanoelectronic systems. Currently, testing can account for up to 60% of the total costs of production for an ASIC - even for 250nm CMOS (SIA 1999), and this figure will become worse at higher densities. Run-time self-test regimes will therefore be increasingly important in the nanocomputer domain.

Wiring Delay

At a basic level, the wiring delay problem is simple to articulate: as interconnection width and thickness decrease, resistance per unit length increases, while as interconnections become denser (and oxide layers thinner), capacitance also tends to increase.

For example, if the RC delay of a 1mm metal line in 0.5mm technology is 15ps then at 100nm (in the same materials) the delay would be 340ps (Sylvester and Keutzer 2001). Ho, Mai and Horowitz (2001) have performed a detailed analysis of the performance of wires in scaled technologies and have identified two distinct characteristics.

For short connections (those that tend to dominate current chip wiring) the ratio of local interconnection delay to gate delay stays very close to unity - i.e. interconnection delay closely tracks gate delay with scaling. For metal interconnections, this will be true down to approximately 10nm at which point the simple resistance relationship breaks down and the resistance increases due to quantum effects.

On the other hand, global wiring tends to increase in length with increasing levels of integration, implying that the interconnection delay of these wires will increase relative to the basic gate delay. Sylvester and Keutzer (2001) conclude that the scaling of global wires will not be sustainable beyond about 180nm due to the rising RC delays of scaled-dimension conductors.

However, as interconnect delay will be tolerably small in blocks of 50 – 100K gates, they argue for hierarchical implementation methodologies based on macro-blocks of this size.

In addition, at future gigahertz operating speeds, signal "time-of-flight" and attenuation will become significant limiting factors. As both of these depend on the dielectric constant of the propagation material, solving them will require significant changes to processing technology.

For example, Ohmi et al (2001) have developed processes that use a gas-isolated, high-k gate dielectric, metal-gate, metal-substrate SOI scheme with thermally conducting through-holes to reduce temperature variations and increase interconnection reliability. These complex, aggressive fabrication schemes contrast markedly with the intrinsic self-assembly mechanisms proposed by Merkle (1996) and others.

Emerging Devices

The problems associated with the scaling of CMOS devices have led to a search for alternative transistor and circuit configurations.

Proposals for silicon-based technologies include silicon-on-insulator single electron devices, resonant-tunnelling devices (RTDs), double layer tunnelling transistors and Schottky Barrier MOSFETS.

Of these, RTDs appear to hold the most promise as a short to medium-term solution although most of the implementations in the literature to date are based on relatively complex heterostructure technologies - predominately based on GaAs. RTDs are inherently fast and have been known and used for more than a decade.

Their negative differential resistance (NDR) characteristics directly support multivalue logic styles (Waho, Chen and Yamamoto 1996) that can result in significantly simpler circuit designs.

RTD circuits are typically based on one or more tunnelling diodes integrated with a conventional FET (Mohan et al 1993).

The main problem with this approach has been the need to match the current of the FET and the peak current of the diode(s) although a recent configuration avoids this problem by surrounding a vertical resonant diode structure with a Schottky control gate (Stock et al 2001). Ultimately, electronic devices may simply cease to be an option at the scale of 1 or 2 nm.

A number of molecular based technologies have been suggested as potential alternatives as well as some computing architectures that might exploit them. There have even been suggestions for nanomechanical devices - somewhat reminiscent of Shannon's original (1949) relay logic as well as computational DNA systems. Finally, semiconductor behaviour has recently been demonstrated within very narrow carbon nanotube (fullerene) based structures.

Nanotube technology may eventually support the construction of non-volatile RAM and logic functions at integration levels approaching 1012 elements/cm2, operating frequencies in excess of 100GHz and, as electron flow can be ballistic in short nanotube wires, supporting current densities in the order of 109A/cm2 - well above the figure that would vaporize traditional interconnect metals. Simple logical operations with nanotubes have just been demonstrated.

Rueckes et al (2000) have built a bistable bit and designs for electromechanical logic and memory have been proposed. Further, high band-gap materials such as boron nitride may also offer interesting nanotube building blocks capable of working at significantly higher temperatures than carbon. Although some of these emerging device technologies have been demonstrated in the laboratory, it is not at all clear which of them have the potential for robust, highspeed operation at room temperature - at least in the near future.

NANOCOMPUTER ARCHITECTURE CANDIDATES

To date, architecture research has responded to the opportunities and challenges offered by device scaling in two ways.

The first approach simply increases existing machine resources - more or larger caches; more on-chip processors, often including local DRAM (Kozyrakis and Patterson 1998), direct multi-threading support (i.e. exploiting parallelism between concurrently running tasks rather than within a single algorithm) and other similar techniques.

While being effective for some applications, these can quickly run into all of the physical limitations outlined previously, especially the wire-length problems that can result in unacceptable memory and I/O latency although the 50 to 100K-gate hierarchical design blocks suggested by Sylvester and Keutzer (2001) are certainly large enough to contain a small RISC processor or other quite sophisticated processing elements.

The second approach uses modular and hierarchical architectures to improve the performance of traditional single-thread architectures. Table 1, reproduced from Fountain et al (1998), compares the three main classes of parallel architectures in terms of characteristics applicable to the nanocomputer domain.

They conclude that highly regular, locally connected, peripherally interfaced, data-parallel architectures offer a good match to the characteristics of nanoelectronic devices. However, it is worth noting that data-parallel architectures represent only a small portion of the interesting problems in computer architecture and are a poor match for most general purpose computing problems. Future computer architectures may well be market application driven (Ronen et al 2001), with the characteristics of each market segment resulting in its own optimised parallel microarchitecture. Ronen et al, like Durbeck and Macias, clearly rule out the possibility of today's high-end microprocessor being tomorrow's low-power/low-cost solution.

Table: A Comparison of Three Parallel Architecture Classes

Parameter	Data	Function	Neural
Degree of parallelism	High	Low	high
Processor Complexity	Low	High	Medium
Interconnect Density	Low	High	high
Amount of Interfacing	Low	High	low
Extensibility	High	Low	low

Quantum Cellular Array Architectures

Cellular Arrays (CAs) have been known and studied for almost 40 years (von Neumann 1966). Their architecture is based on the replication of identical processing elements with nearest neighbour connection. The fundamental idea behind the operation of Quantum Cellular Automata (QCA) devices is that the energy state of a suitable assembly of electrons, initially in a specific ground state, will alter as a result of changed boundary conditions. Lent et al (1993) and more recently Porod (1998) have proposed specific realizations of this idea using twoelectron cells composed of four quantum dots in which the polarization of one cell induces a polarization in a neighbouring cell through Coulomb interaction in a very non-linear fashion.

If left alone, the two electrons will seek the configuration corresponding to the ground state of the cell by tunnelling ("hopping") between the dots. Lent et al have demonstrated that AND gates, OR gates, and inverters can be constructed and interconnected. Fountain et al (1998) comment that circuits built from QCA elements would form extremely coherent computing systems, although some concerns remain about their theoretical validity, and the optimum implementation of memory. As the coulomb interactions in QCA are based on a small number of electrons (as low as one) they tend to be swamped by thermal noise unless they are operated at very low temperatures (in the milliKelvin range).

This will very likely prevent them having a serious impact on the mainstream computing domain. An interesting variation on the QCA - based on magnetism - is described by (Cowburn and Welland 2000). In the Magnetic QCA (MQCA), networks of interacting submicron magnetic dots are used to perform logic operations and propagate information. As MQCA energies are in the order of 1eV they will work well at room temperature. Cowburn and Welland suggest that MQCA technology may eventually offer active device densities in the order of 2.5 x 1011/cm2 with a power-delay product that is 104 times less than current CMOS.

Synthetic Neural Systems

Synthetic Neural Network (SNN) systems, also called artificial neural networks, connectionist networks, or parallel distributed processing networks, are concerned with the synthesis, design, fabrication, training and analysis of neuromorphic (i.e. brain-inspired) electronic systems . These systems achieve high performance via the adaptive interconnection of simple switching elements that process information in parallel. Arrays of simple neural processing elements show features such as association, fault tolerance and self-organisation. However, while the complexity of neural processing is low, the interconnection density is high so there is still a question as to their applicability in the nanocomputer domain.

So far, most of the work in neural networks relates to static networks - classifier systems or associative networks (Glösekötter, Pacha and Goser 1998) that learn to map data by modifying their internal configuration. For example, in addition to employing QCA cells to encode binary information, Porod (1998) has proposed an analogue Quantum-Dot Cellular Neural Network (QCNN) in which each cell is described by appropriate state variables, and the dynamics of the whole array is given by the dynamics governing each cell plus the influence exerted by its neighbours.

The alternative approaches - time dependent, biologically inspired networks that process data using a dynamical systems approach - exhibit more interesting emergent behaviour. They require vast numbers of devices to implement but these are likely to be available in the nanocomputing domain. However, as in all CNN systems, each neural node has to be connected to at least 10 to 100 synapses for useful computation, so it is questionable whether the low drive capability of nanoelectronic devices will be suitable building blocks for these systems.

Locally Connected Machines

A common example of regular, locally connected, dataparallel architectures is the Single Instruction Multiple Data machine. SIMD machines exploit the inherent data parallelism in many algorithms - especially those targeting signal and image processing. Fountain et al (1998) identify the characteristics that may make the SIMD topology suited to nanocomputer architecture as: · a regular and repetitive structure;

- *Local* connections between all system elements;

- All external connections made at the array edge;
- The existence of feasible strategies for fault tolerance.

However, SIMD architecture still suffer from two major problems - global instruction issue as well as global control and clock signals. Global clocking is required by SIMD machines not only to drive each individual (synchronous) element but also to manage inter-element synchronisation. It is clear from the analysis of Fountain et al (1998) that the interconnection costs of SIMD in the nano-domain are very high - with the majority of the die area in their experiments being taken up by control signal distribution. Numerous asynchronous design techniques (e.g. Hauck 1995) have been proposed to overcome the need for a global clock in SIMD machines.

While it is still unclear whether, in practice, these asynchronous techniques actually offer improved performance, they are at least as good as the conventional synchronous approach and may offer the only means to overcome global communication constraints in the nanocomputer domain. The same considerations appear to constrain other multiprocessor architectures such as MIMD. Crawley (1997) has performed a series of experiments on various MIMD architectures and concluded that inter-processor communications will be limited by the availability of wider metal tracks on upper layers (called "fat" wiring by Crawley). The tradeoff here is between track resistance (and therefore delay) and interconnection density. Crawley also notes that more complex computational structures such as carry look-ahead begin to lose their advantages over simpler and smaller structures once wiring delays are factored in.

Propagated Instruction Processor

The Propagated Instruction Processor was proposed by Fountain (1997) as a way of avoiding the interconnection problem in SIMD arising from its global instruction flow characteristics. In the PIP architecture, instructions are pipelined in a horizontal direction such that the single-bit functional units can operate simultaneously on multiple algorithms. The technique shares many of the characteristics of SIMD, pipelined processors and systolic arrays. One of the primary advantages of the architecture is its completely local interconnection scheme that results in high performance on selected applications.

However, the architecture is still basically SIMD and thus will work best with algorithms from which significant data parallelism can be extracted - e.g. Fountain's examples of point-wise 1-bit AND of two images, an 8- bit local median filter, 32-bit point-wise floating point division and an 8-bit global matrix multiplication.

In addition, the fault tolerance of the PIP may ultimately depend of an ability to bypass faulty processors without upsetting the timing relationship between propagating instructions - something that has not been reported to date.

Merged Processor/Memory Systems - IRAM and RAW

The structure and performance of memory chips are becoming a liability to computer architecture. There are two basic problems: firstly the so-called "memory wall" (or gap) resulting from a divergence in the relative speed of processor and DRAM that is growing at 50% per year .

Secondly, while DRAM size is increasing by 60% per year, its fundamental organisation – a single DRAM chip with a single access port - is becoming increasingly difficult to use effectively. This observation has led to the development of a number of merged memory/processor architectures. Two notable examples of this approach are the Intelligent RAM (IRAM) system and the Reconfigurable Architecture Workstation (RAW). The IRAM system merges processing and memory onto a single chip.

The objective is to lower memory latency, increase memory bandwidth, and at the same time improve energy efficiency. The IRAM scheme revives the vector architecture originally found in supercomputers and implements it by merging at least 16MB of DRAM, a 64-bit two-way superscalar processor core with caches, variable width vector units, and a high-performance memory switch onto a single chip. The RAW microprocessor chip comprises a set of replicated tiles, each tile containing a simple RISC like processor, a small amount of configurable logic, and a portion of memory for instructions and data.

Each tile has an associated programmable switch which connects the tiles in a wide-channel point-to-point interconnect.

The compiler statically schedules multiple streams of computations, with one programme counter per tile. The interconnect provides register-to-register communication with very low latency and can also be statically scheduled. The compiler is thus able to schedule instruction-level parallelism across the tiles and exploit the large number of registers and memory ports.

Reconfigurable and Defect Tolerant Hardware

Reconfigurable hardware can be used in a number of ways: to provide reconfigurable functional units within a host processor: as a reconfigurable coprocessor unit; as an attached reconfigurable processor in a multiprocessor system; or as a loosely coupled external standalone processing unit. One of the primary variations between these architectures is the degree of coupling (if any) with a host microprocessor.

For example, the OneChip architecture integrates a Reconfigurable Functional Unit (RFU) into the pipeline of a superscalar Reduced Instruction Set Computer (RISC). The reconfigurable logic appears as a set of Programmable Function Units that operate in parallel with the standard processor.

The Berkeley hybrid MIPS architecture, Garp, includes a reconfigurable coprocessor that shares a single memory hierarchy with the standard processor, while the Chimaera system integrates reconfigurable logic into the host processor itself with direct access to the host's register file.

Reconfigurable Logic and FPGAs

When FPGAs were first introduced they were primarily

considered to be just another form of (mask programmed) gate array - albeit without the large start-up costs and lead times. Since then FPGAs have moved beyond the simple implementation of digital (glue) logic and into general_purpose computation.

Although offering flexibility and the ability to optimise an architecture for a particular application, programmable logic tends to be inefficient at implementing certain types of operations, such as loop and branch control. In addition, there is a perception that fine-grained architectures (those with path widths of one or two bits) exhibit high routing overheads and poor routability.

It is probably true that fieldprogrammable gate arrays (FPGAs) will always be slower and less dense than the equivalent function in full custom, standard cell or mask programmed gate arrays as the configuration nodes take up significant space as well as adding extra capacitance and resistance (and thus delay) to the signal lines. The challenge will be to find new organisations that optimise the balance between reconfigurability and performance. FPGAs exhibit a range of block granularity. Very finegrained logic blocks have been applied to bit level manipulation of data in applications such as encryption image processing and filters while coarse-grained architectures are primarily used for wordwidth datapath circuits.

Again, the tradeoff is between flexibility and performance - as coarse-grained blocks are optimized for large computations, they can be faster and smaller overall than a set of smaller cells connected to form the same type of structure.

However, they also tend to be less flexible, forcing the application to be adapted to the architecture just as for conventional processors. New techniques are required that maintain the flexibility of the FPGA structure while minimising the effects of its configuration overheads. In particular, the serial reconfiguration mechanisms of current FPGAs will clearly not scale indefinitely - a device with 1023 programmable elements could take a few million years to configure! One of the few current proposals to directly address this issue is the Cell Matrix architecture (Macias 1999).

Defect Tolerant Hardware

As identified previously, defect tolerant architectures will be the only way to economically build computing systems with hundreds of billions of devices because any system using nanoscale components will contain significant numbers of defects.One example of an existing defect tolerant custom configurable system is the Teramac. The basic idea was to build a system out of cheap but imperfect components (FPGAs in this case), find the defects and configure the available good resources using software. The high routability of the Teramac is based on the availability of excessive interconnections - due to its "fat-tree" routing configuration. However, it is possible that current methods for detecting defects such as those used in Teramac will not scale to devices with 1010 configuration bits.

Thus, novel parallel defect mapping techniques will need to be developed - most probably built-in, and coupled with self-configuration mechanisms of the type suggested by Macias (1999) or Gericota et al (2001). "Embryonics" is a biologically inspired scheme that aims to produce highly robust integrated circuits with self-repair and self-replication properties. In this case, the reconfiguration algorithm is performed on-chip in the form of an "artificial genome" containing the basic configuration of the cell. Its fault tolerance relies on fault detection and location via built-in self-test plus an ability to bypass faulty cells and to substitute spare cells in their place. However, the simplistic system employed – substituting entire columns of cells if just one cell is faulty - has too many limitations to scale successfully, not the least of which is the need to estimate the number of standby logic cells that might be required in a typical implementation.

While the various demonstration systems have their limitations, they do illustrate that it is possible to build a computer system that contains defective components as long as there is sufficient communication bandwidth to support the discovery and use of working components plus the capacity to perform such a rearrangement of working components. An ability to perform self-test will be critical. It is possible that the most important components in nanocomputer architecture might turn out to be its configuration switches and controls.

NANOCOMPUTER ARCHITECTURE

Having surveyed the current challenges and opportunities in the nanoelectronic domain, it is now possible to make some predictions about likely characteristics of future nanocomputer architectures.

As has been seen, these characteristics lead to: the need for extremely localised interconnect; the use of homogenous arrays that are able to support heterogenous processing structures; the ability to exploit parallelism at multiple levels (e.g. instruction level, multi-threaded etc.); a requirement for dynamic reconfigurability with low reconfiguration overheads (in both space and time) as well as defect and/or fault tolerance - at both the commissioning/configuration stage and at run-time.

RECONFIGURING THE MEMORY GAP

Although the fabrication of RAM and digital logic are completely separated at present, and there is a vast and expensive infrastructure supporting both, the functions of logic and memory must eventually merge if the increasing gap in performance between the two is to be overcome. Non-volatility will be the key. When DRAM is finally superseded by non-volatile memory, it will be possible to envisage a computing system in which all storage – disk, main memory and caches – merges into the processing mesh.

Figure illustrates one reason why this would be a good idea. In the memory hierarchy of a conventional processor, it is possible for code and data items to be duplicated in more than five places in the system (e.g. disk, disk cache, memory, memory cache(s), registers). It is fairly easy to argue that this is not a good use of the available machine resources (Flynn 1999).

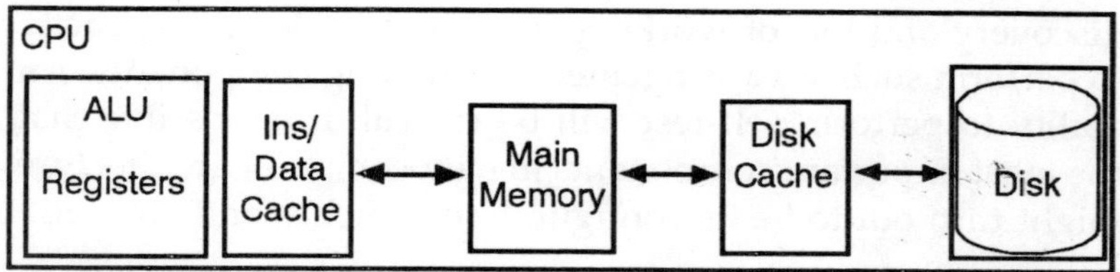

Fig. Conventional Memory Hierarchy

At present, the two main contenders for non-volatile technology are floating-gate structures and magnetics. The roadmap for non-volatile Magnetic RAM (MRAM) shows it reaching a density of 1Gbit by 2006 (Inomata 2001) and "nano-magnetic" technology (Cowburn and Welland 2000) may eventually support densities of 1012 bits.

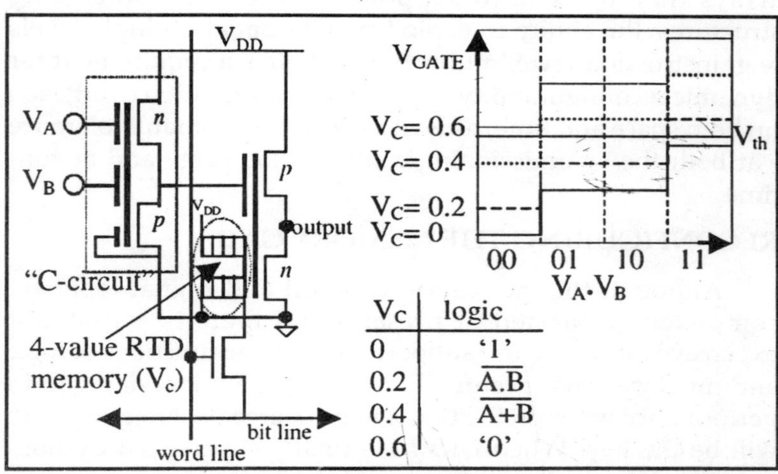

Fig. Reconfigurable Threshold Logic Cell

Although floating-gate devices have been under development for around 30 years, they are unlikely to reach the same densities as magnetic based systems. Figure illustrates an example of a non-volatile reconfigurable logic cell that merges the variable threshold nMOS logic of Kotani, Shibata and Ohmi (1992) with the multi-valued memory cell of Wei and Lin (1992).

In this cell, the initial "c-circuit" acts as a simple D/A

converter, producing a voltage that is proportional to the two input values. The RTD-based 4- valued memory is used to adjust the offset of this voltage, shown as VGATE in Figure thereby reconfiguring the function of the (two-input) logic gate.

The low-overhead reconfigurability offered by this type of circuit – or by alternatives such as nano-magnetics – may eventually support the creation of a merged memory/ processing structure, in which the idea of "mass storage" is replaced by "mass reconfiguration" as programme and data become indistinguishable from the processing mesh.

Coding Around the Defects

As outlined previously, all nanocomputer systems will contain faulty components. Defect/fault tolerance supporting the ability to detect and avoid defects at both the commissioning/ configuration stage and at run-time, will therefore be of critical importance. Forshaw et al (2001) have shown that it is theoretically possible to produce working systems with defect rates as high as 10-5 to 10-4 if reconfigurable techniques are used to bypass the defects.

Existing static fault mapping techniques (such as are used in hard disk systems, for example) may represent a good starting point, but it is likely that built-in self test (BIST) will be necessary to maintain system integrity in the presence of soft-errors and noise. There have been some initial studies into how to optimally configure BIST in an extremely large cellular array but no general solutions have been developed as yet.

"Grain Wars"

At least in the short term, the outcome of the coarse-grain vs. fine-grain argument is difficult to predict as there are strong arguments for both styles. Eventually, however, all nanocomputer architectures will be formed from arrays of simple cells with highly localised interconnect. This will be an inevitable outcome of shrinking geometries as devices evolve towards molecular and quantum/ single electron technologies. At present, the tendency towards course-grained architectures (e.g. multiple CPU blocks, ALU arrays etc.) is being driven by the high overheads imposed by reconfiguration techniques in devices such as FPGAs.

If this can be reduced, for example by the use of multi-value techniques such as was illustrated previously, then finegrained structures offer a much more general solution to the creation of flexible computing platforms.

A Processor for Every Process

It appears, then, that the ultimate computing fabric will be an homogenous, fine-grained, non-volatile, fault tolerant, reconfigurable, processing array, exhibiting adjacent or nearest neighbour interconnect only and supporting heterogenous structures that are derived by compiling a HLL programme. The processing fabric will be reconfigurable in a way that maximises the system's ability to exploit parallelism - consisting of as many individual processing meshes as are necessary, each configured in an optimal manner for the particular function.

This scheme takes advantage of the future abundance of processing with a scarcity of interconnect. Instead of a large number of constructed programs, we may instead try to store (close to) all possible programs in the device.

In this organisation, programs would be continuously configured within the non-volatile memory/logic - available to respond to an input stimulus by generating an output whenever required. The concept of memory hierarchy would be completely eliminated – if the logic structure is large enough to store and execute all "programs" for that machine.

In the more immediate term, configuration "context switching" will replace the loader of conventional operating systems. As this architecture effectively merges processor logic, RAM and disk into one structure, the only remaining potential performance bottleneck will be the input/output channel. Although I/O bandwidth tends not to be as great a problem as the memory/processor interface, current processors working in domains such as multimedia already have some difficulty maintaining high data throughputs and this will continue to be an issue (for example with 3D multimedia). The challenge will be to develop flexible parallel I/O configurations that will allow the internal processes to operate at peak performance.

Legacy Software

General purpose computing is largely sequential, dominated by control dependencies and tends to rely on dynamic data structures that currently do not map well to array architectures. However a nanocomputer will inherit a vast quantity of legacy software that cannot be ignored (it could be said that the Y2K issue revealed just how extraordinarily long-lived are some types of software). There is no doubt that the translation from a source programme to systems with billions of gates will be an extremely complex task.

But, ironically, the very availability of a large number of gates makes the task easier. In this case, the synthesis process has access to the resources necessary to create all possible computation paths in the "programme" and then simply select the single correct result at the end. This aggressive form of speculation is the basis of the synthesis process for the PipeRench architecture.

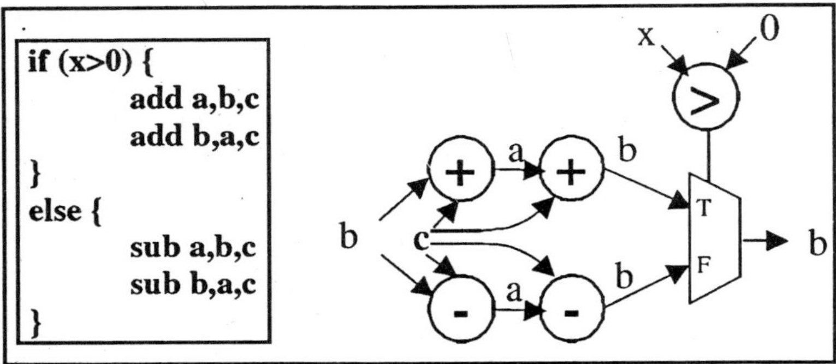

Fig. Processing Graph Fragment

The graph fragment in Figure illustrates this point. All arithmetic functions are duplicated as required, as are the intermediate variables - without concern for the hazards that would occur in a typical pipelined system. Note that in this simplified diagram, no data synchronisation mechanism is shown. A number of ideas have been proposed that would be applicable to the nanocomputer domain, from Asynchronous Wave Pipelines (Hauck, Katoch and Huss 2000) through to gate-level nanopipelined computation using RTDs.

We argue that future nanocomputer architectures will be formed from non-volatile reconfigurable, locallyconnected hardware meshes that merge processing and memory. In this paper, we have highlighted the characteristics of nanoelectronic devices that make this most likely - primarily the severe limitations on the length interconnection lines between devices. It appears that the current trend towards coarse-grained structures may not be supportable in the long term. If the overheads associated with reconfigurability can be reduced or even eliminated, architectures based on fine-grained meshes with rich, local interconnect offer a better match to the characteristics of nanoelectronic devices. Of course, having access to a vast, reconfigurable computing platform is only the first step. The question still remains as to what use such an architecture might be put.

Will it be necessary to own an "Avagadro computer" in order to run WindowsÒ 2030? Moravec (1998) has suggested that, if the power of the human brain is in the synapses connecting neurons, then it would take the equivalent of 1014 instructions/sec. to mimic a brain with an estimated 1013 - 1015 synapses. Might the power of nanocomputer architecture finally release the ghost in the machine?

Chapter 8

Towards a Reconfigurable Nanocomputer Platform

Reconfigurable architectures are of great interest to system designers because they offer a way of achieving power and performance efficiency by matching algorithmic constructs with the appropriate architectures. The traditional approach to reconfiguration, in FPGAs for example, has been to build separate areas of programmable logic gates and interconnection blocks from transistors and to manage the two resources more-or-less separately during the synthesis process. Much of the work on constructing reconfigurable platforms has therefore been directed towards answering the questions "how much of each and in what form?".

This is largely because reconfiguration is expensive - it increases the area and power requirements, while reducing the performance of a system compared to purpose-built solutions[1]. It also places constraints on the solution space accessible to the synthesis process that are unique to the particular platform, reducing the generality of system solutions. Achieving optimum performance in a reconfigurable system by striking a balance between logic and interconnect is often difficult.

The continued scaling of devices into the deep-sub-micron region will have a number of consequences for reconfigurable systems. For example, it appears likely that only arrays of simple cells with highly localised interconnect will be available. This will be the inevitable result of the reduced fanout, power handling capacity, gain and reliability of DSM devices. Thus it is likely that the boundaries between logic and interconnect will become increasingly difficult to identify.

On the other hand, as a result of scaling novel circuit techniques are becoming available: vertically integrated and thin-film based transistors, heterogenous resonant tunnelling devices, single- electron quantum dots and spin filter devices to name just a few. This paper addresses the basic question: how can low-overhead reconfigurable systems be built with nanoscale technology? It is proposed that the answer to this question is really the same as that for achieving high fabrication density - to exploit the "3rd dimension" wherever available.

The use of aggressive vertical integration is resulting in some extremely compact circuits and 3D or spatially stacked techniques promise to increase this significantly. However, this is not the only "dimension" available. High density, high functionality configurable building blocks can also be achieved by simultaneously exploiting a number of available physical variables (voltage, current,threshold voltage; magnetic field, electron spin, etc).

Three plausible configurable building blocks will be described that illustrate how this might be achieved. In Section, an example is presented of how transistor logic scaled into the nanometer domain can be combined with RTD-based multi-valued logic to support fine-grained configurability.

A second example, in Section , exploits thecharacteristics of phase transition thin-films and vertical integration to reduce the size of the cell and at the same time support non-volatile operation.

In Section , a cell based on magnetic tunnellingtechnology is described that exploits a number of physical variables to merge logic and configuration into a single space. The fabrication of these buildingblocks will not be described in exhaustive detail. Rather, the intention is to explore how the particular techniques might offer high functionality within dense structures. Finally, a general direction that computing applications might take on such a homogenous non-volatile processing mesh.

A MULTI-VALUED SRAM BASED PLATFORM

This example mesh brings together the Resonant Tunnelling based multi-valued SRAM described in and more recently in, the metal-insulator tunnel transistor (MITT) and a some of the 3D interconnection ideas described, for example, in. The resultingstructure forms a large planar logic mesh of simple (2-input) logic gates in two of its dimensions, which can be configured into a interconnected logic structure via a vertically stacked (multi-valued) RTD-based RAM array.

Although their characteristics are similar, theMITT device operates quite differently to aBack Gatesubstrate Insulatorstandard FET. The gate voltage modulates the tunnelbarrier thickness, changing the Fowler-Nordheim tunnelling currents that flow through the metal/insulator junction. In the proposed device, a second (back) gate has been added so that the tunnelling current becomes a function of both. A number of dual-gate transistors have been described but none based on MITT technology[2]. This idea has the advantage that the all-metal devices are compatible with current fabrication processes and can be easily extended to multi-layer 3D topologies.

Accessing the two gates independently allows the upper gate to be used to set the operating point of the transistor - effectively setting the threshold voltage of the lower gate that provides the logic connection. By setting the transistor bias points appropriately, a logic gate can be built that exhibits three operating regions always on (V_{G1}>0.1), always off (V_{G1}<-0.4) and inverting operation with a logic swing of 0.4V (-0.2 <V_{G1}< -0.1). The top-gate voltage V_{G1} will become the reconfiguration voltage for the homogenous logic mesh.

The 3-state memory cell proposed by Wei et al and more recently by van der Wagt provides a useful mechanism for generating the three threshold voltages required. The basic memory cell is redrawn in Figure along with its (idealised) load-line curves.

To merge this into the processing mesh will involvematching the V_{G1} values required to set the MITT into its three operating domains with the RTD tunnelling voltages

$V_{1\text{-}3}$ which are set, in turn, by adjusting the (AlAs) barrier thicknesses of bothRTDs.

The top gate connections form the contact point to the RTD memory cells.

In the configuration shown, memory cells are paired with each pair setting the configuration for the two-input NOR gate below. The cell layout depicted in Figure is based on a similar layout described in and illustrates how the heterojunction RTDs might be assembled on a substrate grown over an insulator layer containing the MITTs.

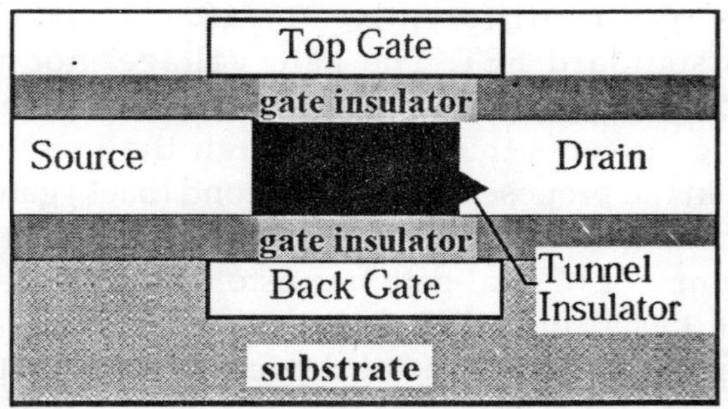

Fig. Proposed Double-Gate Metal Insulator Tunnel Transistor

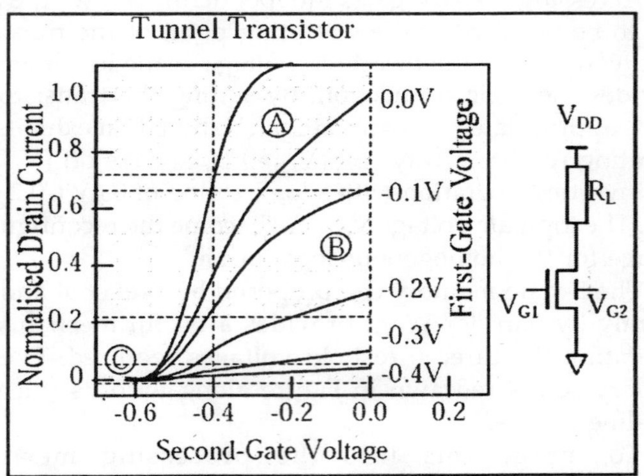

Fig. Typical Dual Gate Transfer Characteristics

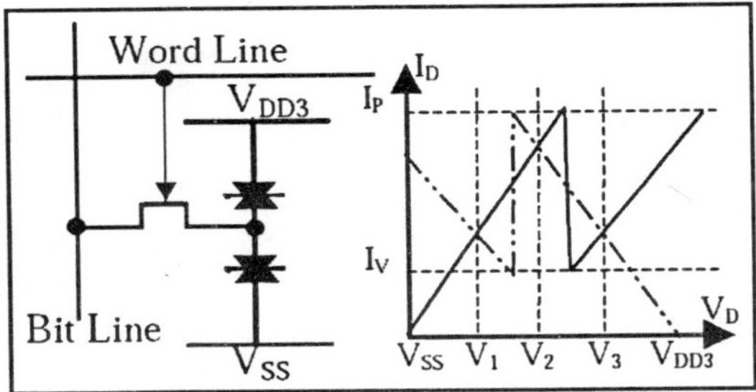

Fig. RTD-Based SRAM Cell

A major problem with RTDs is the need to closelycontrol the growth of the tunnelling layers as thetunnelling current depends exponentially on the tunnel barrier thicknesses. For large scale integrationIII/V technology appears to be the most promising, although a number of significant processing challenges remain. At present the integrationdensity is limited by the minimum lateral featuresizes for RTD devices (about 1 μm) and theiroperating currents (peak currents are about 5mA).The Nanotechnology Roadmap predicts that theywill scale to about 200nm by 2006 and around 50nmby 2012. At this point, these RTD-based memorycells could be built with dimensions down to about300nm x 100nm, and the MITT based logic cellsshould scale to ~10nm gate lengths. This represents adensity in the order of 3 x 10^9 cells/cm^2, or theequivalent of about 10^9 2-input gates/cm^2.

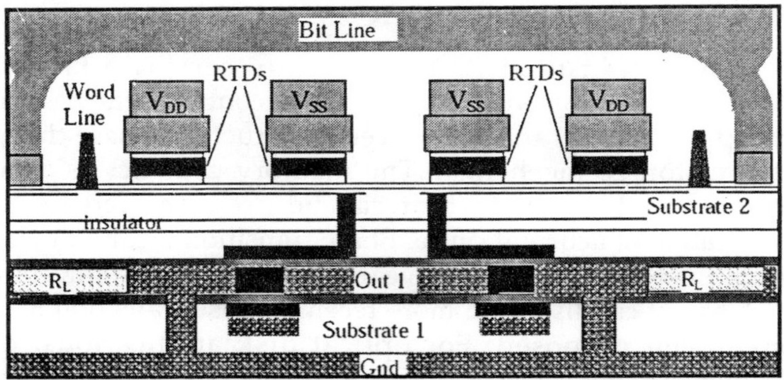

Fig. Vertically Integrated Logic Cell Structure

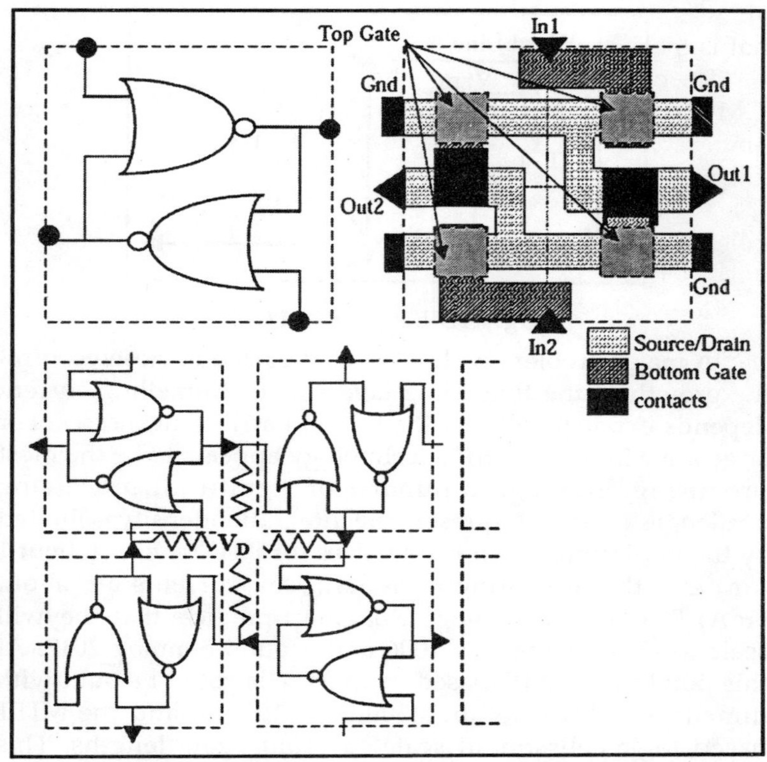

Fig. An Example Logic Cell, Layout and Tiling

PHASE TRANSITION DEVICES

Phase transition RAMs generally comprise layersof chalcogenide semiconductor thin-films (containingat least one group-VI element) and use a reversiblestructural phase change as the storage mechanism.The memory cells act as a fast programmable resistorthat can be programmed by the application of acurrent pulse. Phase transition technology has beenin use for a number of years in non-volatileapplications such as optical disks and more recentlyphase-transition RAM have been proposed. For optical disk applications, the difference in optical characteristics between the crystalline and amorphous states defines the logic state of an individual cell, whereas for RAM applications it is the resistance of the film that is measured.

The chalcogenide alloy material is compatible withcurrent (CMOS) logic fabrication, adding only 2 - 4late, low temperature mask steps to a standardprocess. The technology scales well as smallermemory cells use less power during the phasechange, operate at lower voltages, are less prone toelectric field damage and therefore have much greaterwrite/ erase lifetimes. Stable operation has alreadybeen reported with phase transition areas down to20nm in diameter and multi-state storage viaprogramming to intermediate resistance values hasbeen demonstrated.

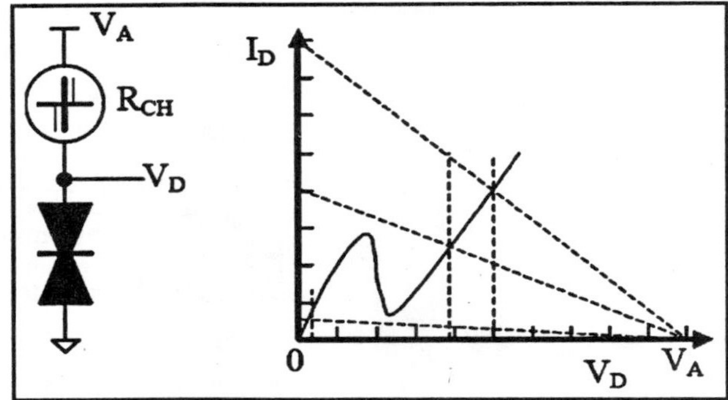

Fig. Multi-Valued RTD/Phase-Transition Memory Cell

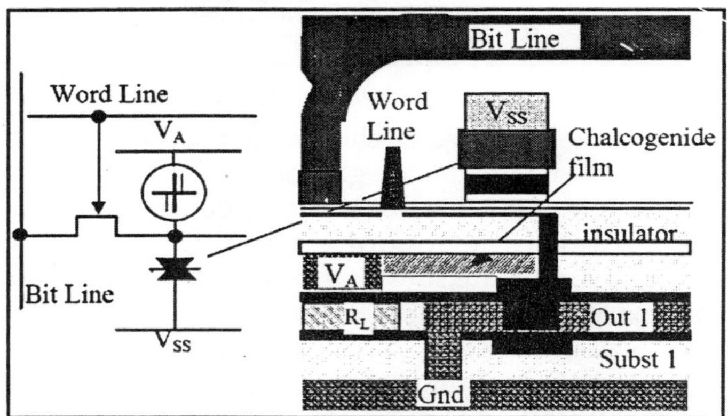

Fig. RTD/Phase-Transition Memory Cell Configuration

By using a chalcogenide resistor as the load for aresonant tunnelling diode, it is possible tocreate a smaller multi-valued RAM cell that is alsonon-volatile. As shown, the variable resistance of thephase-change film is used to establish the threeoperating points as before. The design of the cell willinvolve matching the operating resistances of thechalcogenide material to the V-I characteristic of theRTD, and at the same time minimising the staticpower dissipation of the configuration cell. This willneed to be held at less than 1nW per cell if thepotential maximum mesh densities are to be realisedin practice.

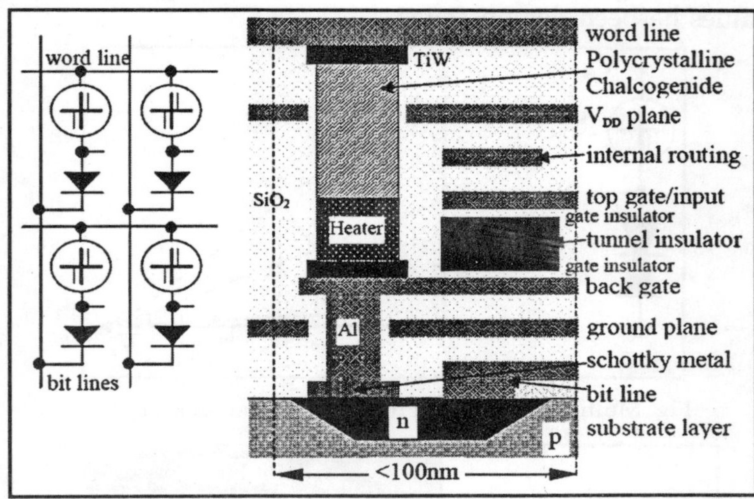

Fig. PROM Style Cell Layout

A possible 3D layout shown in Figure merges the chalcogenide film with the top gate of the MITT. The resulting structure is compact, with a likely final cell size under 150nm square. However, by adopting a conventional (PROM) memory array structure,substituting a schottky diode for the RTD andexpanding vertically the cell can be madeeven more (laterally) compact.

In this configuration(shown in this case on a standard silicon substrate)the schottky diode acts both as current isolation forthe memory array and to set the reference point forthe logic array. Although a silicon junction isprobably simpler to fabricate, GaAs might be a betterchoice as it broadens the choice of metals that mightbe used to form the schottky junction. Improvedmetal-to-metal diode junction technology may evenallow the semiconductor substrate to be eliminated entirely.

The cell layout as drawn assumes that thechalcogenide film and its resistive heating elementcan be fabricated as a narrow (e.g. 20-30nm) verticalpillar. The objective is to control the operatingresistance of the film and thus manage the staticpower dissipation of the configuration cell. In Figure8 the MITT source and drain connections areperpendicular to the page and the vertical connectionsto ground etc. have been omitted. Seven layers ofinterconnect metal are used, including the power andground planes necessary to reduce inductance andcross-talk in the bit and word lines.

During normal operation, the word line is held at the reference voltage (V_A in Figure)and the bit line at V_{SS} (or ground). Programming a particular bit involves setting the bit/word voltages such that only the bit to be programmed sees a voltage above the transition threshold across the chalcogenide film. As the individual reference/MITT cells can be connected by abutment, with adjacent cells sharing a pull-up device (e.g. a thin-film resistor), this layout will be very compact - gate density could reach the equivalent of 4×10^9 reconfigurable 2-input gates per cm^2.

It is also likely that deep trench techniquessimilar to those used in the S-SGT DRAM device,with the non-volatile chalcogenide film replacing thestorage capacitance, could reduce the footprint evenfurther.

It is worth noting that this density is achieved at thecost of system level flexibility. Individual bits are nolonger programmable while the logic mesh isoperating as the bias conditions of the entire meshneeds to be disturbed to programme just one bit. Thiscould be partially managed by partitioning the arrayinto separately programmable areas (which might benecessary in any case, due to electrical loading-considerations) but this would simply transform the problem into one of deciding the optimum size forthese partitions. Both of these first two examples share acharacteristic in common with the majority of currentreconfigurable devices such as FPGAs in that theconfiguration mechanism and the logic mesh occupycompletely separate spaces. In this case, unless themesh layout is augmented to allow for logic plane I/O(for example, by incorporating a separate word/bit read mechanism on additional layers), it would be constrained to the edge of the chip. In the final example that follows, logic and reconfiguration are merged into the same physical space.

A NANO-MAGNETIC PLATFORM

Thin-film magnetic technology seems set tochallenge dynamic and non-volatile memories based on conventional semiconductors. For example, Cowburn and Welland have suggested that nanomagnetic technology (Magnetic QCAs, in their case) may eventually support non-volatile active device densities in the order of 2.5 x $10^{11}/cm^2$ with a power- delay product that is 10^4 times less than current CMOS.

Magnetic memories are being developed based on the giant magneto-resistive (GMR) effect using devices such as spin-valves and pseudo-spin-valves in which patterned magnetic multi-layers change resistance with changing magnetic moments between the layers.

Devices typically comprise two layers of ferromagnetic material separated by a nonmagnetic insulator and are "on" (lower resistance) if the direction of magnetization in the two adjacentlayers is the same and "off" (higher resistance) if it isin opposite directions. The layers must be narrowerthan the magnetic domain wall for the effect toemerge, but it remains stable and reproducible withdecreasing geometries. However, only a verylimited effect has been achieved to date with reportedDR values ranging between 18% and 40% at roomtemperature. This is adequate for non-volatile RAM applications in which the resulting small voltage changes can easily be detected. GMR-based logichas been proposed but its operating currents arestill much too large for high density integration.

An alternative magnetoresistive tunnel device hasbeen suggested that has a potential GMR manytime larger than this - possibly as high as 10^5.

Thedouble spin filter tunnel junction (DSFTJ) reversesthe usual structure of spin devices in that the tunnelbarrier material is formed from two different layersthat are insulating but are magnetic and exhibitunequal coercivities

The coercivity of onelayer is chosen to be very large so that it remains"pinned" (unable to be changed), while the other hasa smaller value that the direction of its magnetizationcan be changed more easily.

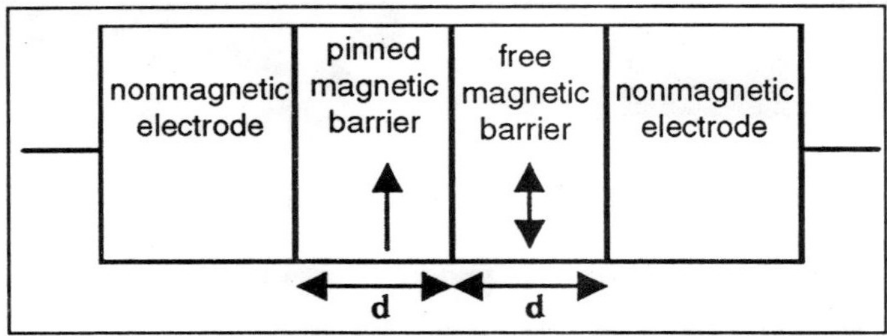

Fig. Double Spin Filter Junction (from)

In DSFTJ layout proposed in Figure 10, the thinmagnetic barriers are formed on the surface of aconductive pillar. The resistance is varied betweenthe inner pillar and the multiple outer conductors thatprovide nearest neighbour connection. In this case asquare lattice is shown but other layouts (e.g.hexagonal) would be equally possible.

While much of the enabling technology exists,there are still a number of significant technicalchallenges to be overcome to make the device work.High density pillars of metallic and/ or magneticmaterials have been grown for applications such assuch as e-beam lithography, RIE and plating or nano-imprint lithography have resulted in the production ofpillars and rings of material down to 10nm diameterson 40nm centers with aspect ratios of 10:1.However, it is likely to be difficult to find materialswith just the right combinations of small band gapand large exchange splitting, that exhibit appropriatehysteresis curves and can be grown in thin, pinhole-free layers on a lattice-matched conductor. It hasbeen demonstrated that ferrites can be grownepitaxially on $SrTiO_3$ so this would represent agood starting point in the search for suitablematerials.

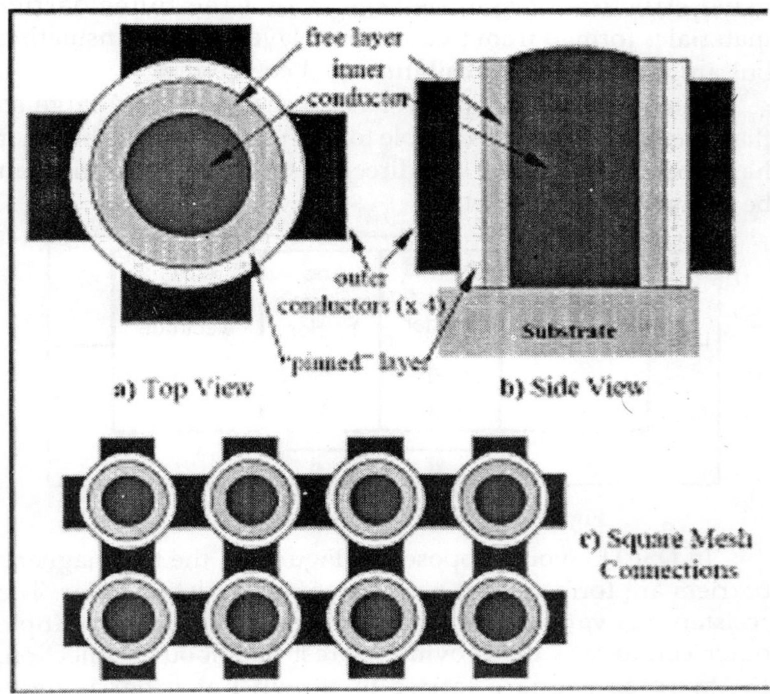

Fig. The "Bullseye" Double Tunnel Junction

The model for the DSFTJ developed by Worledgeand Geballe predicts that its magnetoresistancewill be an exponential function of the layerthicknesses so the device becomes more sensitive asthe thickness of the barrier layers is increased.Unfortunately, its resistance is also an exponentialfunction of barrier thickness making it difficult tocontrol the summing currents between devices.

Theprogramming current must be able to be set at a valuesufficient to saturate the free layer without switchingthe pinned layer which implies that either the twolayers must start with widely differing coercivities, orthe variation across the mesh needs to be very closelycontrolled. A careful tradeoff will need to be made betweenthe thickness (and therefore the GMR), the tunnelresistance, the coercivity of the two layers and theavailable programming current.

Other importantissues will include magnetostatic interactions notonly between the layers but possibly between theadjacent nanomagnets within the array, theeffect of diameter and thickness on the hysteresiscurves and domain duplication between layers.

The adjacent devices create a logic mesh in which resistive interconnections may be created by manipulating (programming) the magnetic fields within the spin junctions. In the network of Figure, currents are summed into the non-linearity of the RTD.

This logic array is functionally identical to the collection of resistively coupled quantum dots proposed by Roychowdhury et al except that now the connectivity is programmable and non- volatile. Just as in Roychowdhury's proposal, achieving uni-directional logic operation is a potential problem with this array.

Processing would start with the RTD substratefollowed by the patterning of the top RTD contacts.The contacts would then form both the mask for themesa etch stage of the RTD and the seed points for the growth of the contact pillars. Finally, afterepitaxial deposition of the magnetic layers over the pillars and patterning the interconnect metal, a layer of oxide would be grown and planarized ready to take the upper layer of active devices.

A straightforward way to address and programme the individual junctions is to arrange them in a RAM-like array by integrating an isolation transistor into the cell. A Vertical Tunnelling Transistor (VTT) would be particularly suited to this application as it can be formed by a selectively patterned epitaxial metallisation of a SOI substrate grown over the magnetic array and the overall device is likely to be scalable to less than 1nm square, while maintaining high current density.

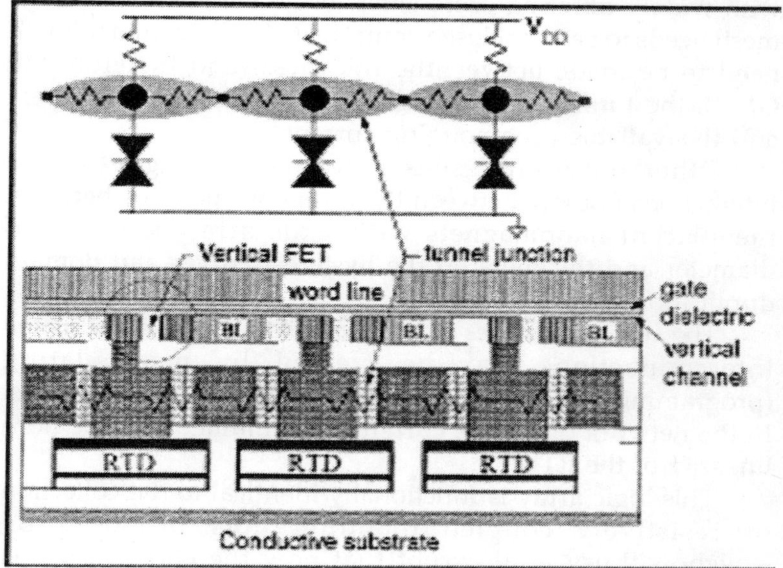

Fig. Magnetic Logic Element

In this case, the VTT is connected, via the lower side of its source, to the central pillar of the DSFTJ and acts both to supply the programming currents and to read the voltage level of the result. The top side of the source is pulled up to V_{DD} via a thin-film resistor or similar (not shown). Only one current pulse[4] is needed to configure devices to on or off. In additional, the isolation transistor provides a convenient way of applying a global signal (such as a multiphase clock signal) to the logic mesh.

This vertically integrated approach could eventually support tunnel junction densities in excess of 2 x $10^{11}/cm^2$, although how this translates into logic gates is complicated by a number of issues.

Firstly, while it takes a minimum of three junctions to form a two-input logic gate, each independent logic block must be surrounded by a palisade of isolationcells. The actual number of these isolation cells will depend the logic optimisation and the actual placement topology and has an optimistic lower bound of $4mn^{1/2}$ (for m logic blocks each with n junctions) -

which tends to favour fewer, largerblocks as a layout strategy. Isolation also needs to beprovided for faulty gates. This includes not only hardfailures (i.e. short-circuits) but devices that fail logicand/or parametric tests - exhibiting low current orlonger than acceptable delay times. While it is clearthat reconfiguration is a very powerful technique formaintaining reliability in a logic network, mostanalyses appear to be based on the assumption thatfaults are randomly distributed and their effects arelocalised. However, the blocking caused by a clusterof faults in an area may cause significant "flow-on"effects for that could conceivably cause a largenumber of otherwise good cells to becomeunavailable to the place and route process.

As logical inversion is not possible in this type of system, all logic variables and their inverse must be maintained right through the system, potentially doubling the number of junction cells used for a particular function. Finally, as for standard MRAM devices, on-chip space is required by support circuits such as current-drive and voltage detection, address generation etc.

RECONFIGURABLE SPATIAL COMPUTING

The decision about how to perform a particularcomputational task has traditionally involved strikinga balance between hardware and software - betweenthe spatial and the temporal domains. At oneextreme, full custom hardware offers parallel spatialexecution with very little flexibility or generality. Atthe other, software implementations exploit a more-or-less "general-purpose" platform to provide aflexible but lower performance solution.

To reduce the performance gap between these twoextremes, complex hierarchical memory schemeshave emerged that try to hide the cost of moving codeand data items from one place to another in aprocessor system. However, it is clear that thereturns from the increasingly complex structures usedfor functions such as speculation, reordering, andcaching are rapidly diminishing and mayeventually become counter productive[5]. The move tonanoscale devices will accelerate this trend. The concept of 3D memory has been proposedas a way of bringing memory and processing physically closer together (Figure). If theoverheads associated with reconfigurability can bereduced sufficiently, there is no fundamental obstacleto completely removing this separation of memoryand processing and with it the performance gapbetween the two. This may ultimately lead to acomputing system in which all storage - registers,caches, main memory and even eventually disk - isunified within the processing mesh (Figure).

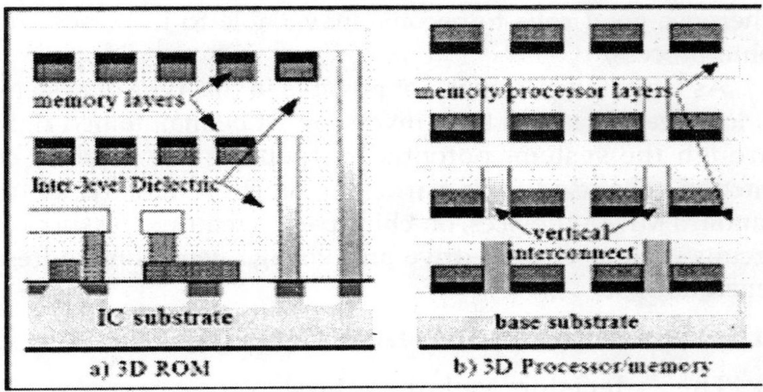

Fig. Comparison Between the 3D-ROM Concept and 3D Merged Processor/memory.

CONCLUSIONS

In this paper, some ideas have been presented about how low-overhead reconfigurable systems might be built to exploit the characteristics of emerging nanoscale devices. Although not all of the enabling technology is currently available, the example circuits presented have demonstrated that it will be possible to exploit a number of alternative "dimensions" - apart from the obvious spatial dimension - to construct compact reconfigurable cells.

Dual-gate transistors are of great interest as theyoffer a way of avoiding short channel effects andsubstrate leakage (in CMOS) as channel length isscaled down. Given that the second gate can bejustified on these grounds, the ability to configure thetransistor's operating region can be added with verylittle additional overhead. All that is required is anefficient means of developing the multi-levelreference voltages and this has been illustrated usingboth RTD and phase change chalcogenide devices.

A high-density non-volatile reconfigurable cell hasbeen proposed in which a double junction spin filtertunnel junction is built on a vertical conducting pillarand integrated into a nearest neighbour-connectedmesh. Whether the basic device can actually be builtrelies on the discovery of materials with the rightcombinations of small band gap and large exchangesplitting, that exhibit appropriate hysteresis curvesand can be grown in thin, pinhole-free layers on alattice-matched conductor. The availability of this sort of high-density non-volatile reconfigurable mesh can provide a flexible, defect tolerant platform that will allow novel 3D spatial computer architectures to be explored.

Chapter 9

A Route Toward a Chemically Assembled Electronic Nanocomputer

INTRODUCTION

In the early 1960s Rolf Landauer began to consider the physics of information manipulation. Allmodern computers are Boolean logic, nonreversible machines. Tnis implies that the states '1' and '0'need to be energetically distinct from one another. Furthermore, in a nonreversible machine, informationis lost. Consider, for example, a simple AND logic gate. That gate has two inputs and one output. Thestate of the output depends, of course, on the collective states of the inputs, so that if both inputs arehigh, then the output is high. However, if any or both of the inputs is low, then the output is low. Let'ssay that we do a simple calculation with an AND gate, and we produce a '0' on the output. Any one ofthree different input states could have given us this output, and we can't tell which one.

Thus, the'nonreversible' term used to describe the computing machine refers to the fact that information is lostduring the calculation. In another example, say that we carry out the following addition on our calculation:5 + 5. The calculator generates the number '10', and now we show that number to a colleague. Ourcolleague can't tell if your question was 5 + 5 = ?, or 2 × 5 = ?, or 3 + 7 = ?, etc. Thus, in this type ofmachine, the question is lost, and that has entropic consequences. Landauer demonstrated that theminimum energy required to carry out a single bit operation, considering both entropic and enthalpiccosts, was k_BTln2.

This minimum energy, however, requires that the bit manipulation takes placevery slowly. As any chemist knows, the faster a chemical reaction takes place, the larger the DG ofreaction. This is also true for information manipulation, as both processes must obey the laws of thermodynamics. Feynman considered the bit operations to take place in a finite amount of time (at aparticular clock frequency), and he demonstrated that the power consumption of this (slightly morerealistic) process was $P = nk_BTdn^2c^{-1}$. In this equation, P is the power consumption (in watts), n isthe number of parallel operations, n is the clock speed, c is the speed of light, and d is a transmissiondistance.

It is not clear just what the appropriate metric for optimizing a machine is, but it is reasonablystraightforward to demonstrate that 10^{18} bit operations/second at a power consumption of 1 W is notunrealistic for a massively parallel machine. Feynman's equation implies several things. First, note thatthe power consumption increases linearly with the number of parallel operations, but quadratically withthe clock speed! Thus, parallel computers are much more energy-efficient than serial machines. Second,d should be small, and so an energy-optimized computer is going to be characterized by small (nanometer)length scales. Finally, if the energy difference between a '0' and a '1' is k_BT times some factor, then aquantum-state switch is required to achieve such efficiency. In the talk preceding mine, Stan Williamsdiscussed the difficulties involved in manufacturing such a machine with current microelectronicstechnology.

First, complementary metal oxide semiconductor (CMOS) devices are not quantum-stateswitches, but instead are solid-state switches. Second, true quantum-state switch operation under ambienttemperature requires length scales below 10 nm. The costs associated with bringing the CMOS (top-down) manufacturing technology down this length scale are truly unrealistic! This means that if one isto build a machine that operates near the thermodynamic limit for computation, then that machine isgoing to use a bottom-up manufacturing approach (chemistry), and the best quantum-state switchesobtainable via chemical routes (or perhaps any other routes, for that matter) are molecules. In fact,molecules are quite natural quantum-state switches.

At first glance, chemical manufacturing and microprocesser fabrication appear to be incommensurate with one another, for at least three reasons.

- First, modern microprocessors are perfect and complicated systems, whereas anything that ischemically fabricated tends to be either crystalline (or amorphous), and is characterized by im- perfections due to reaction yields, lattice defects, etc. Thus, one challenge for developing a bot- tom-up manufacturing strategy is to find a way to extract perfect complexity from defective order.
- Second, any machine (even a human brain) has a very finite number of inputs (the keyboard, for example) and outputs (the screen), but a much larger number of computational resources (100 Mbytes RAM, Pentium Pro II, etc.). In a computing machine, the few inputs/outputs are con- nected to the resources through structures known as multiplexers and demultiplexers. The analo- gous structures for a chemically assembled machine will have to reach down from the micron or millimeter length scales into the nanometer or Angstrom length scales. An architecture for such a multiplexer has not yet been developed, and will not be discussed here.
- Third, any chemically assembled machine is likely to have a vast amount of resources. How doesone take advantage of so many resources, and do so in an efficient manner? In a modern CMOS-based machine, the resources are laid out according to the following: Engineers know that, on asingle cm^2 of Si wafer, the current lithography generation will enable

them to build a certainnumber of wires and switches, amplifiers, etc. Those resources need to be arranged in such afashion that they will be efficiently utilized for a general or specific purpose computing machine. Thus, an algorithm is written that can take all of these resources and attempt to find a ground-statesolution for arranging them together. This algorithm is run several times, until everyone is satis-fied that a reasonably optimal solution has been found. Here's the catch. This problem is knownas an 'NP-hard' problem, which means that the difficulty in finding a solution scales exponen-tially with the number of resources. The more familiar 'traveling salesman' problem is similar.Thus, trying to find an optimal solution for arranging 10^{23} resources will take longer than the ageof the universe! This means that an architecture must be found that makes solving this problemscale linearly with the numbers of resources, rather than exponentially.

THE ARCHITECTURE

The set of problems outlined above appears to be fairly significant, and we haven't yet decided onwhether we need a two- or three-dimensional architecture, or even what the quantum-state switches willbe! I will discuss the issue of dimensionality first, since all other solutions will hinge on that. In the early 1960s, Richard Rent of IBM developed a phenomenological rule known today Rent's rule. A simpleRent's rule calculation is shown in Figure. The point of Rent's rule is that within any portion of any givencircuit there are a certain numbers of degrees of freedom. In order to take advantage of all those degreesof freedom, then a given fraction of the wires need to be devoted to input/output (I/O). In a perfectlydesigned system (that may do nothing interesting other than be perfectly designed), Rent's rule says thatfor n degrees of freedom, there must be $2n^{1/2}$ I/O wires. Here, the ½ exponent is called the Rent's ruleexponent. When human design is taken into account, the Rent's rule exponent becomes somethingcloser to 0.6 or 0.65. If the circuit has any defects, then additional wires must be added to route aroundthose defects, and the Rent's rule exponent becomes 0.7 or higher. This is for a 2D circuit. For a 3Dcircuit, perfect design dictates a Rent's rule exponent of 2/3, and human design, defect, etc. will makethat number 0.75-0.8 or so. The problem with a 3D circuit is that the surface area of a volume onlyscales, in the best case, to the 2/3 power of the volume, and so there is just enough surface area to utilizeall of the resources. Once human design, defects, etc. are taken into account, there are not enough wires.This implies that either the system must be utilized in a very inefficient manner (remember, the scalingis in the exponent), or that the efficient use of 3D architectures is just not possible for a Boolean logic,reversible machine. Thus, the issue of 3D versus 2D was already settled by Richard Rent 40 years ago!We need a 2D architecture for our fundamental circuits. It turns out that it is possible to network in 3D,although not in a manner that is space-filling. However, the issue of networking is beyond the scopeof this paper.

Now we come back to the issue of chemical assembly. We already stated that any chemically assembled system is likely to be crystalline and defective. The fact that it is crystalline, and that we are going to have a 2D architecture, implies that we need an architectural solution that tiles in 2D. A few

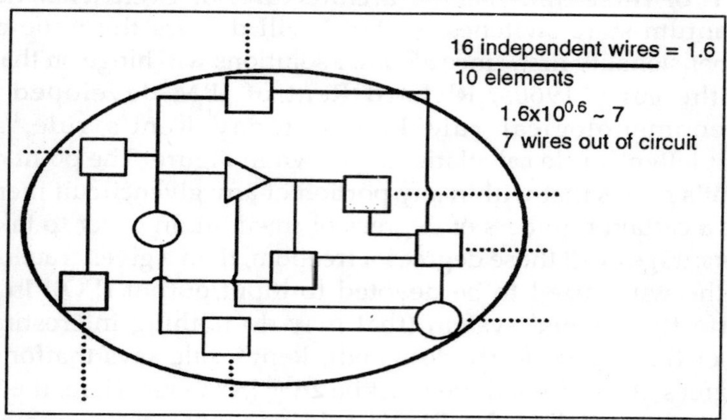

Fig. A Graphical Depiction of Rent's Rule.

This drawing represents a circuit containing logic gates (squares, circles, and triangles), and wires. There are 10 gates and 16 wires, including the input and output wires. Note that not all of the wires are independent, and only the independent ones are counted. As shown in the figure, this circuit has a Rent's rule exponent of approximately 0.7 years ago, I, along with two collaborators from Hewlett Packard Labs, Stan Williams and Phil Kuekes, began trying to reconcile the issues of chemistry with computer manufacturing, and we used a machine known as Teramac as a model for our discussion. Teramac is a massively parallel, custom-configurable supercomputer built at HP in the early 90s by Kuekes, Greg Snider, and others. Teramac also was an extremely defective computer—it contained nearly a quarter million hardware defects, any one of which would completely disable a Pentium processor. Nevertheless, when Teramac was running, it was capable of 10^{12} logical operations/second, and it only had a clock speed of 1 MHz!

Teramac turned out to be very important in developing our thoughts for how to build a CAEN, and so it is worth spending some time discussing this machine. In Teramac, the fundamental device was a configurable bit, properly known as a field-programmable gate array (FPGA). The fundamental circuit was a square lattice of FPGAs, known as a crossbar, and the fundamental architecture was known as a fat tree. It turns out that the concept of a crossbar of configurable bits actually solves the problem of how to deal with a defective manufacturing technology that produces only crystalline structures. The concept of the fat tree makes the problem of utilizing a vast number of resources one that scales roughly linearly with the numbers of devices, rather than exponentially. Fat trees are exten-sions of earlier concepts, one example of which are known as Banyan networks.

Teramac was a chameleon of a computer. As built, it was nothing but wires and switches, with no ability to compute. Kuekes and his colleagues knew ahead of time that their machine would be riddled with defects, and so they designed the machine with this in mind, meaning that they designed circuits with Rent's rule exponents of 0.7 and higher. They wrote algorithms that sent electrical signals through-out the entire machine, similar to how some computer viruses work. Those algorithms established two data sets: the good resources and the defective ones. Once these data bases were in hand, it was possible to devise a way to efficiently utilize the good resources in order to build a computer. Remember, all they had were resources, but they had a lot of them, and so they could imagine using these resources to fabricate virtually any known type of computer. Rather than physically moving the good resources around on the circuit boards to assemble the computer, the entire assembly was done through software.

Thus, the logical structure of the machine was downloaded upon the good hardware resources. This 'building' of the computer took a lot of time and was therefore expensive. However, it was done by a second computer that served as a tutor to Teramac. The software, coupled with the high Rent's rule exponent used in the circuit design, allowed for defective resources to be routed around.

Configuring the resources of Teramac into a computing machine was made vastly easier because of the fat tree architecture. A fat tree is a hierarchical arrangement of the resources, similar to a family tree in which kids, parents, grandparents, etc. are all at different levels. The difference between a fat tree and a family tree is that the fat tree has a breadth, not just a width. One example of a fat tree is the system of roads, boulevards, and freeways in Los Angeles. An example of how a fat tree is used is shown in figure. In this analogy, there are thin lines (neighborhood streets), thicker solid lines (boulevards), and even thicker triple striped lines (freeways).

In a fat tree, all of the neighborhood streets are connected to each other and to the boulevards that intersect the neighborhood. All of the boulevards are connected to each other and to the intersecting freeways, and so on. Finding a pathway from one house to another house in the same neighborhood (short-distance communication) is relatively easy, and all pathways are relatively the same. Getting from neighborhood to neighborhood (intermediate distance communica-tion) is relatively easy through the use of boulevards. Finally, getting from one part of the city to another (long-distance communication) is also relatively easy through the use of freeways.

Many choices are nearly equivalent. It is not critical to find the shortest distance route, as a solution reasonably close to the 'ground-state' solution can always be found. Teramac had a 5-level fat tree, and any CAEN is likely to need a dozen or so levels to deal with the vast numbers of resources. Note that, without a map, finding an efficient route between the two points of Figure would be nontrivial. It is even possible that, given an order of magnitude more time to find a solution (but no map), no good solution would be found.

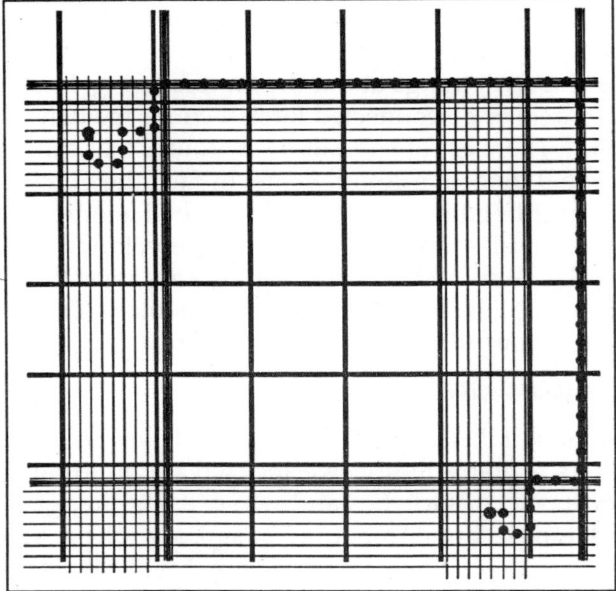

Fig. A fat tree Architecture modeled after a Street map of an ideal Town.

Only a few of the 'minor roads' (thin grey lines) are shown, while all of the major boulevards (thick grey lines) and all of the freeways (triple grey lines) are shown.

The object is to find a good way to connect the two dots, and one possible pathway (black dotted lines) is shown. The assumption that the freeways are better for long distance communication has been made, and the boulevards are better for intermediate distance communication. Note that many solutions are possible, but it is relatively easy to a nearly optimal one.

This illustrates the usefulness of retaining, but not building in, the 3rd dimension. It is only upon looking down onto these resources from the 3rd dimension that one finds an efficient pathway between any two points. If this map had contained twice the numbers of resources, it would have only taken approximately twice as long to find an optimal pathway between two points. Such an experiment was run on the Teramac resources. Thus, the fat tree goes far toward solving the issue of scaling.

Note several things about the fat tree of Fig. First, broken streets (wires) or disconnected inter-sections (switches) can be dealt with. There are always alternate routes that can be found which avoid defective components. In fact, the efficiency of a fat tree architecture probably does not become se-verely hampered until somewhere near the percolation threshold. Second, this is an architecture that appears crystalline (albeit with differing length scales for streets, boulevards, freeways, etc.), and it does tile in two dimensions.

THE HARDWARE

We have taken a stepwise approach toward building a CAEN. As a first attempt, we focused on building a single 'neighborhood' of switches and wires, using lithographically defined wires, but molecular-based switches. Since the switches are the new component here, I will discuss them in some detail. As stated above, the Teramac machine utilized FPGAs as configurable bits. The logical structure of a single FPGA is shown in Figure. Note that this is not a simple structure! It contains 4 wires—two data lines (d) and two address lines (a), plus a memory bit (m) and a switch (s). It contains a total of 6 transistors, 4 of which are used in the inverting circuit to keep the device nonvolatile. Even though it might be possible to find molecular-based components that can reproduce all of the properties of a single FPGA, it is probably not worth it. What is much simpler, however, is to come up with a molecular-based system that

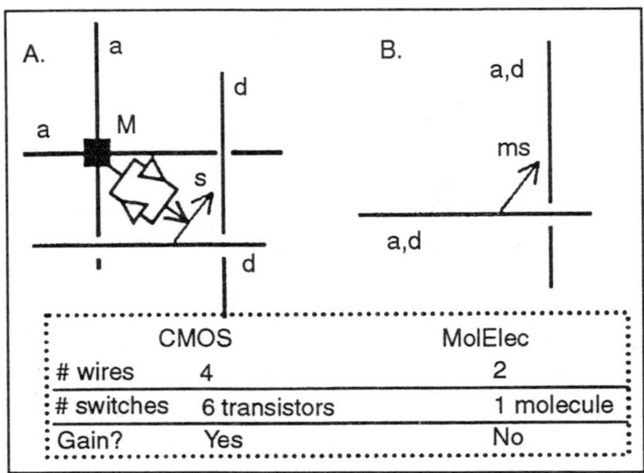

	CMOS	MolElec
# wires	4	2
# switches	6 transistors	1 molecule
Gain?	Yes	No

Fig. A direct Comparison with a field-programmable gate array (A) with a Molecular Electronic Switching Junction (B).

reproduces certain of the most critical qualities of an FPGA: namely, high on/off switching ratios; nonvolatility; and the ability to electronically configure the switch, and then read its state.

An advantage of molecules is that they are voltage-addressable. This means that different volt-ages can be utilized to carry out different functions. Imagine a molecule that is sandwiched between two conducting electrodes. This molecule is essentially a tunneling barrier, since the only way that current is going to get from one electrode to the other is via tunneling through the molecule. Since tunneling is a quantum mechanical process, virtually anything that is done to the molecule is going to have an expo-nential effect on the tunneling rate between the two electrodes. Let's say that we prepare this device, and we measure the junction resistance at 0 applied volts to be x Ohms. We then oxidize the molecule at +1 V. When we look again at the junction resistance at 0 V, it is now y Ohms. If we have found the right molecule, then x and y are widely separated values. Since we read the device at 0 V, the memory is nonvolatile. In addition, we have a device in which the two wires are essentially address lines when used at +1V, but they are data lines when used at 0 V. Finally, this device is based on a molecular property, and so should scale down to single molecule/nanowire junctions.

One point that we have treated rather casually in the above discussion is the following: If we read the device at 0 V, and write it by oxidizing the molecule at +1 V, then what is to keep the device from reducing back to the unoxidized form when we lower the voltage back to 0 V? That would then give us the exact same junction resistance that we had before we oxidized the junction. It turns out that this will most certainly be a problem, and there exist two possible solutions. First, we can make the oxidation process irreversible, but in that case we can only use the switch one time. While this does not make the device useless, a much more interesting device would be one that was reconfigureable. Thus, the most attractive option is to find a way to build hysteretic character into the molecular junction voltammograms.

This second option would make the operation of our devices very similar to cross-point ferromagnetic memory devices that are currently being developed by a number of companies. In those devices, known as Mag-RAM devices, there are giant magnetoresistive (GMR) materials that act as 'spin valves.' If the GMR material is electrically poled in a particular direction, then the junction resistance of the crosspoint element is reduced. This is similar to a magnetization vs. field poling process, such as that used in hard drive data storage. The difference is that it is electrically addressable, rather than addressed with a read-head.

There are various classes of molecules that might serve as attractive candidates for these devices. In particular, however, molecules that have a large amount of structural reorganization that accompa-nies a redox process are chief considerations, since such molecules should be characterized by activa-tion barriers to the actual redox process. Such activation barriers are necessary for generating hysteric voltammograms. One set of candidates are the molecular-mechanical compounds known as

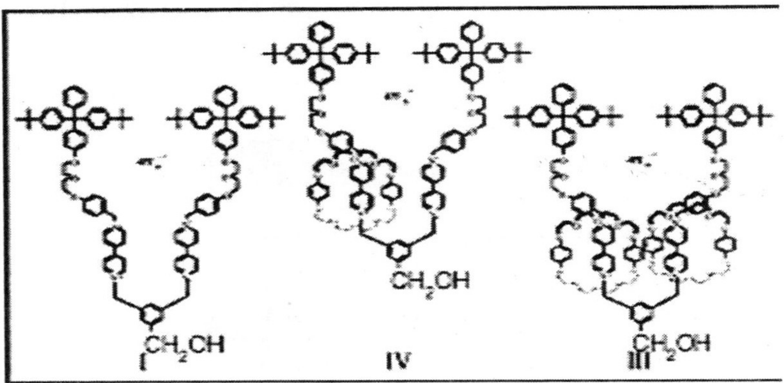

Fig. The three Molecular Compounds used in this study. Only (B) and (C) are Properly Rotaxanes, since only those two Molecules Contain Mechanically Interlocked Components. rotaxanes, catenanes, and pseudo-rotaxanes.

These classes of molecules had been developed by Fraser Stoddart and his group over a nearly 20-year period starting in 1981. Fraser's group had even shown (in the solution-phase) electrochemically addressable switching and even simple logic operations from certain of these systems. Equally important was that his group had demonstrated that these various classes of molecular compounds were highly modular. This means that it might be possible to have components that switch, components that aid assembly, and components that add chemical stability—all in the same molecular compound.

The molecular compounds that we began to work with, in collaboration with Stoddart's group, are shown in Figure. These molecular compounds come from the class of molecules mentioned above that are known as rotaxanes. Rotaxanes consist of two or more components mechanically inter-locked with one another: one of the components is dumbbell-shaped, and the remaining components are rings that become trapped on the dumbbell component, encircling part of the dumbbell.

In the unique class of rotaxanes used here, the dumbbell component contains two bipyridinium units. Only the two compounds of Fig. 4 that actually contain the rings are proper rotaxanes. The encircling ring(s) is (are) bis-para-phenylene-34-crown-10.

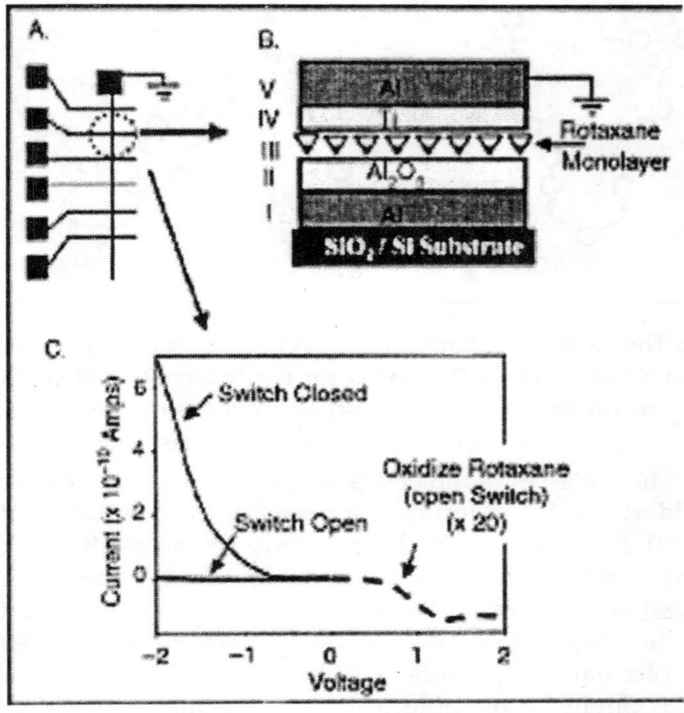

Fig. A diagram of the linear array Arrangement of Devices utilized to make Wired Logic-gates.

Based on both their structure and their solution-phase properties, these rotaxanes are not expected to exhibit bi-stable switching. This means that they are not candidates for a reconfigurable switch. However, all of the compounds irreversibly oxidize, which means that they are candidates for singly configurable switches. The reduction process on the unoxidized systems is reversible. The point of making these devices was to investigate the following questions:

- Can we successfully incorporate monolayers of rotaxane molecules into a sandwich-type device within a crossbar structure?
- Can we make a solid state, singly configurable switch from these molecules?
- If we can make such a switch, can we correlate those switching properties with solution-phase voltammetry measurements? And
- If we can make such switches in high yield, can we electronically configure several of these devices together to make wired-logic devices?

The answers to all these questions turned out to be 'yes', and much of that work was recently published. Only some of the highlights of that work will be presented here. In Figure we present the actual structure of the devices that were fabricated. A cross-sectional view depicting the assembly of the devices. A SiO_2 substrate is patterned with a an aluminum wire (I) that is characterized by approximately 1 nm of Al_2O_3 native oxide (II). On top of this bottom electrode a single monolayer of rotaxanes molecules (III) is transferred as a Langmuir-Blodgett film. The second electrode is deposited as (IV) 50 nm of Ti metal followed by (V) 1 mm of Al metal.

The Fermi levels of both electrodes are essentially identical, so that all diode behaviour observed in the devices originates from the molecules. (C). The switching properties of these molecular junctions is shown in this plot. The device, as originally prepared, represents the switch in its closed state, and it is read at negative (reducing voltages). Reduction of this junction is a reversible process, and the device may be read many times without significant degradation. The switch is 'opened' by writing the device at positive voltages. This serves to irreversibly oxidize the molecule. When the opened switch is read at negative biases, the current through the junction is now reduced by a factor of approximately 10^2.

the operation of one of these switches. Similar devices had been previously fabricated and character-ized, and modeled in my group. The structure shown in Fig. is obviously a linear array arrangement of devices, not a crossbar. However, crossbars were fabricated as well, and devices in those structures performed in a manner similar to the linear arrays. It was also possible to correlate the electrochemical properties of the individual solid-state devices with solution-phase voltammetry. Finally, a critical question that we were seeking to answer related to the issue of the electrical response of the switches—i.e., the difference between 'open' and 'closed'. The Mag-RAM devices previously discussed are made in a similar fashion (as a crosspoint memory), but those junctions show only a very small difference between a '1' state and a '0' state (about 15%). While that is sufficient for Mag RAM devices to be utilized for memory applications, it is not sufficient to use them for logic. In a wiredlogic gate, several devices are strung together, and the entire circuit is dissipative.

In Fig. we observe a factor of approximately 10^2 between a '0' and '1' state, and thus logic devices could be fabricated from our switching junctions. The truth table of a wired AND gate is shown in Fig. In this gate, the junction labeled 'L' was an open switch, while the two junctions labeled 'A' and 'B' were open

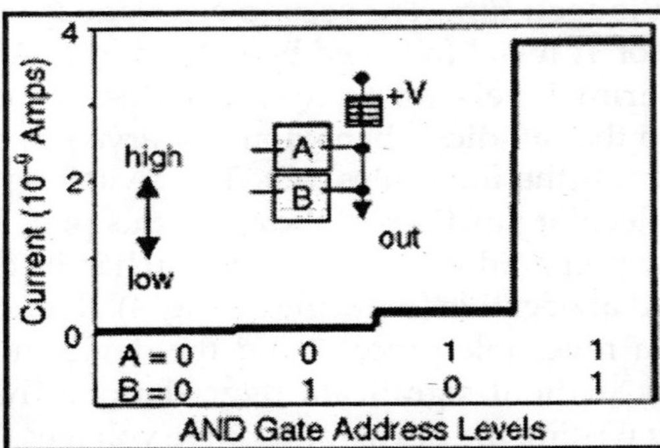

Fig. Experimentally Measured Truth table of a wired-AND gate constructed from these molecular junction arrays. The hi/lo levels of the truth table are separated by a factor of 15 switches.

The truth table was measured at -1.5V, while the 'L' switch was opened at about +1 V. In our recent paper we were also able to demonstrate that the concept of electronically configuring these devices into logic gates was defect-tolerant.

Chapter 10

Nanostructures and Bio-Inspired Computer Engineering

INTRODUCTION

Nanoelectronics is the development of electronic devices with dimensions between one and ten nanometers. If Moore's law continues to hold, it could be possible to construct Avogadro Computer, i.e. an architecture that efficiently employs 10^{23} elements to perform computation, sometimes about 2020. Shorter-time prediction says that chips will contain 10^{10} transistors and will operate at 10-15 GHz in 2012. It is not clear how, but two major groups of problems will have to be solved: (1) How to build Avogadro Computer physically, and (2) how to manage such an amount of resources effectively. While the first group covers technological problems like scaling limits, power management, reliability limits etc., the second group of problems (which are relevant for this paper) deals with computational architectures for nanocomputing.

We are interested in nanocomputers because they should provide higher degree of parallelism than current computer technologies are able; massive parallelism is a good way to handle computational bottleneck. The second key feature is that a lot of resources, which can be utilized to implement adaptive and fault-tolerant devices, will be available.

In this paper we emphasize the application of bio-inspired principles to the nanocomputing elements design and to the design of computational models for nanocomputers. We would like to benefit from adaptation, self-assembling, self-repair, self-replication and other features that are observable in living beings. These features have already entered into (software and hardware) computer engineering in recent years. However, their importance will grow in future. In particular we will show how evolutionary design could be applied in nanotechnology. Bioinspired computational models will be presented as hot candidates for the design of nanocomputers.

EVOLUTIONARY DESIGN

Natural computing is a new term that has been established in computer science and computer engineering. It covers the following fields: computational intelligence (soft computing), molecular computing, quantum computing and membrane computing.

Computational intelligence (mainly the areas like evolutionary algorithms, cellular automata, artificial neural networks and algorithms based on the immune system principles) was utilized in hardware design in which new fields such as evolvable hardware, embryonic electronics and immunological electronics have appeared in recent years. The methods of computational intelligence are employed in such cases in which the conventional approaches lead to poor results.

In general we need in nanotechnology to inexpensively arrange atoms in useful patterns permitted by physical laws. The evolutionary algorithm is a widely acceptable population-based search method that can be utilized in optimization or in the design from a scratch. The designer is to encode the problem into the chromosome, define operators over chromosomes (e.g. crossover, mutation) and specify a measure (fitness function) to define quality of every candidate solution. In other words, (s)he has to specify what has to be designed instead of how it has be designed.

For instance, the evolutionary algorithm was employed for the geometry optimization (energy minimization) of clusters and nanoparticles, where the interactions between atoms, ions or molecules are described by a variety of potential energy functions (force fields). These approaches typically use software simulators in order to evaluate candidate solutions. In case of the evolutionary circuit design we can mention digital as well as analog circuits evolved from a scratch in the field of evolvable hardware. The evolutionary design enables to explore the design space efficiently and thus to discover really innovative digital as well as analog circuits far beyond the scope of conventional engineering design methods. It is important that evolution is free to exploit physical, chemical as well as electrical properties of a "target platform" when the evaluation is performed in the physical device. In case of the circuit design, researchers usually work with the physical Field Programmable Gate/Transistor Arrays (FPGA) and the simulators are not used.

Linden demonstrated, that it is possible to evolve electromechanical systems in-situ, i.e. without software simulator. The evolutionary algorithm was employed to find physical construction of the antenna in the reconfigurable antenna array. Similarly to FPGAs, the approach can be extended to the evolutionary molecular and nanostructures design in a Field Programmable Matter Array (FPMA). The idea behind the concept of FPMA is that applied voltages may induce physical changes that interact in unexpected way with other distant voltage-inducted configurations in a rich physical substrate. In other words, it should theoretically be possible to perform the evolution directly in materio if the platform is configurable in some way. Miller and Downing reviewed the promising technology in.

We have very good experience with the evolutionary optimization as well as with the evolutionary design at The Faculty of Information Technology of Brno University of Technology (FIT). We are able to evolve relatively complex digital circuits which cannot be designed using conventional design techniques and which operate better than conventionally designed circuits. It seems natural to apply this technique at the molecular level to build nanostructures, for instance, nanogates (basic elements of digital nanocircuits). We are looking for a physical experimental platform or a suitable software simulator right now.

COMPUTATIONAL MODELS ORANOTECHNOLOGY

During the nanocomputer design process the technological constrains will probably allow only local interaction of uniform simple computation cells since limitations on the length interconnection lines are primary problems.

Hence such a computer has to be based in fact on the cellular automaton model. Furthermore, it is unclear how to reach reasonable reliability for a 10^{23} element device. If the same reliability as current technology exhibits should be available, then the error-rate must decrease in several orders of magnitude. It seems that such devices will require a lot of redundancy and spare elements to assembly and repair itself autonomously.

A number of biological principles such as self-assembling, self-replication, self-diagnostic and self-repair were introduced in the field known as embryonics. The aim is to develop fault tolerance scenarios sufficient enough to suppress inherent higher defect-ratio of nanocomputers and to introduce adaptability and reliability of the applications.

These features are now simulated using computer models intensively. Current hardware implementations developed for testing and research purposes consist only of a few hundreds of components, far from a 10^{23} element device. Two major platforms exist: (1) Embryonic and POEtic projects at EPFL (Switzerland) & University of York (UK) and (2) Cell Matrix platform (USA).

For instance, Cell Matrix is a hot candidate for the computational model that nanocomputing could benefit from. These platforms allow building the circuits based on the ideas taken from biology: multicellular organization, cellular division, and cellular differentiation. Let us mention that the research at FIT is conducted in this direction too.

On the other hand, the computation can be performed on molecules directly. The best-studied molecules for this purpose to date have been DNA and bacteriorhodopsin. While DNA computing is based on the massive parallelism of DNA strands and Watson-Crick complementarity, the protein bacteriorhodopsin, which contains the light sensitive rhodopsin, supports optical computing.

NANOTECHNOLOGY: INHERENTLY A MULTIDISCIPLINARY FIELD

Nanotechnology is inherently a multidisciplinary field, drawing from chemistry, biology, computer science, and mechanical engineering. We showed that computer engineering could benefit from nanotechnology. However, nanotechnology will benefit from computer engineering in future too: First, Peterson says in: "Dealing with the complexity and system engineering challenges will probably fall to computer scientists, who are perhaps best positioned to coordinate, and even direct, the work for the other specialist needed in molecular technology development. As the open source software community has demonstrated repeatedly, software engineers are the cutting edge of grappling with large, physically dispersed teams who are completing tasks of boggling complexity." Second, computer engineers have developed techniques for the evolutionary design that can be employed to exploit physical and chemical properties of the material. These techniques can be used in all cases where we are able to represent the solution in computer. Third, computer scientists have introduced mathematical models of natural computing (e.g. DNA computing is formalized well). These computational models are used to simulate the real world as well as to determine limits of the approaches.

Chapter 11

Computational Nanotechnology: an Assessment

INTRODUCTION

In 1959, Richard Feynman, a future Nobel Laureate, gave a visionary talk entitled "There's Plenty of Room at the Bottom" on miniaturization to nanometre-scales. Later, the work of Drexler also gave futuristic visions of nanotechnology. Feynman and Drexler's visions inspired many researchers in physics, material science, chemistry, biology and engineering to become nanotechnologists. Their visions were fundamental: since our ancestors made flint axes, we have been improving our technology to bring convenience into our everyday life. Today a computer can be carried with one hand - 40 years ago a computer (hundreds of times slower) was the size of a room. Miniaturization of microprocessors is currently in process at nanometre-scales. Yet, the style of our modern technology is still the same as ancient technology that constructed a refined product from bulk materials. This style is referred to as bulk or top-down technology.

As conventional methods to miniaturise the size of transistors in silicon microprocessor chips will soon reach its limit and the modification of today's top-down technology to produce nanoscale structures is difficult and expensive, a new generation of computer components will be required. Feynman and Drexler proposed a new style of technology, which assembles individual atoms or molecules into a refined product. This Drexler terms molecular technology or bottom up technology. This bottom-up technology could be the answer for the computer industry. Though top-down technology currently remains the choice for constructing mass-produced devices, nanotechnologists are having increasing success in developing bottom-up technology.

There are some concerns regarding emergent bottom-up technology. First, the laws of physics do not always apply at nanometre-scales. The properties of matter at nanometre-scales are governed by a complex combination of classical physics and quantum mechanics.

Nevertheless, bottom-up fabrication methods have been successfully used to make nanotubes and quantum dots. These methods are not yet suitable for building complex electronic devices such as computer processors, not to mention nanoassemblers that can make copies of themselves and work together at a task. Furthermore, and significantly, once knowledge of nanotechnology is advanced and real-world nanoassemblers are realized, they must be properly controllable to prevent any threats to our world.

More recently computer science has become involved in nanotechnology. Such research is wide ranging and includes: software engineering, networking, internet security, image processing, virtual reality, human machine interface, artificial intelligence, and intelligent systems. Most work focuses on the development of research tools. For example, computer graphics and image processing have been used in nanomanipulators that provide researchers an interactive system interface to scanning-probe microscopes, which allow us to investigate and manipulate the surface at atomic scales. In addition, genetic algorithms have been used as a method in automatic system design for molecular nanotechnology.

Computer science offers more opportunities for nanotechnology. Soft Computing techniques such as swarm intelligence, genetic algorithms and cellular automata can enable systems with desirable emergent properties, for example growth, self-repair, and complex networks. Many researchers have successfully applied such techniques to real-world problems including complex control systems in manufacturing plants and air traffic control. With some modifications towards nanotechnology characteristics, these techniques can be applied to control a swarm of a trillion nanoassemblers or nanorobots (once realised). It is anticipated that soft computing methods such as these will overcome concerns about implications of nanotechnology, and prevent the notorious scenario of self-replicating nanorobots multiplying uncontrollably.

This article reviews nanotechnology from different points of view in different research areas. We discuss the development of the field at the present time, and examine some concerns regarding the field. We then focus on the needs and benefits of computer science for nanotechnology, as well as existing and future computer science research for nanotechnology.

DEVELOPMENT IN NANOTECHNOLOGY

To describe Feynman's grand visions that have inspired many researchers in several fields of study, Drexler introduced the term "Nanotechnology" and "Molecular Engineering" in his book, "Engines of Creation". He explored and characterized an extensive view of Feynman's visions in many aspects including potential benefits and possible dangers to humanity. According to the vision, building products with atomic precision by bottom-up technology could offer a dramatic widespread of potential and a decrease in environmental impact which would improve our way of life. A simple example of potential benefits from nanotechnology is that information on papers could be packed into much smaller spaces so that less pollution from discarding those papers would be produced. The aspect that would be directly beneficial to humankind is nanomedicine, which involves medical research at nanoscale. For example, a group of programmable nanorobots that could flow along our bloodstreams without harm to our bodies could be injected to treat our bodies from within.

Nanotechnology has indeed promised a great future for humanity. However, the down side of the technology should not be neglected. Drexler suggested the potential threats to life on Earth of uncontrollably replicating assemblers. In order to prevent any threat to the society, it is crucial that nanotechnology is developed under acceptable standards with regard to ethical and social considerations. Recently, the Foresight Institute, which is a non-profit organisation for nanotechnology, gave version 4.0 of its guidelines as a self-assessment list for research and development in the field of nanotechnology. The Science Media Centre has also produced a document describing nanotechnology for use by the media. Today nanotechnology is gaining public attention. Many companies have been doing research and development in nanotechnology for commercial purposes. The governments of several countries have begun funding for research in this area. This recent development of nanotechnology is described further in the following sections.

Nanomanipulators

One important concept of nanotechnology is building products using bottom-up technology. Instead of sculpting bulk materials into desired products, bottom-up technology suggests a new method that assembles individual atoms into products. The first step to bottom-up technology is to acquire the ability to manipulate individual atoms at the scale of nanometres as desired. Therefore, the development of a nanomanipulator, which is a tool for manipulating nanoscopic materials, is seen by some as being crucial to the progress of nanotechnology.

The first imaging in nanoscale was from the electron microscope developed by M. Knoll and E. Ruska in 1931. Later in 1981, G. Binnig and H. Rohrer invented the scanning tunnelling microscope (STM) that can image individual atoms, and earned the Nobel Prize. The success of the scanning tunneling microscope leads to the development of other scanning probe microscopes (SPM) including the atomic force microscope (AFM). Instead of using lenses like traditional microscopes, all these scanning probe microscopes use a probe to scan atoms over the surface, measure a local property and result the image. Different types of scanning probe devices are designed for different tasks. For example, the STM is only appropriate when the material conducts electricity, while the AFM can work with non-conducting materials.

Apart from resembling a surface at atomic scale into a high-resolution image, scanning probe microscopes can be used to manipulate individual atoms. In 1990, D. M. Eigler of IBM used an STM to precisely place xenon atoms on a nickel plate into the name "IBM. In 1993, W. Robinett and R. S. Williams developed a virtual reality system that allowed user to see and touch atoms via the scanning tunneling microscope. This was the beginning of a nanomanipulator. At the University of North Carolina, another nanomanipulator has been developed in a multi-disciplinary project involving in the collaboration of several departments including computer science, physics and chemistry. This nanomanipulator is a virtual-reality interface to scanning probe devices. Using technology in computer graphics, the features that are faint in the image can be enhanced. The system allows scientists to investigate and manipulate the surface of materials at the atomic scale. As a result, it has led to new discoveries in biology, material science and engineering. For example, scientists have used the nanomanipulator system to examine the mechanical and electrical properties of carbon nanotubes. Nanomanipulators are now commercially available. However, the ability to manipulate individual atoms alone could not yet enable us to build reliable nanomachines, unless the physical principles at nanoscales are comprehended.

Nanofabrication

After scientists have gained the ability to manipulate individual atoms directly, the next step is to manufacture structures at nanometre scale, i.e. structures smaller than 100 nanometres across. In this section, we discuss nanofabrication methods, which can be divided into two categories: top-down methods and bottom-up methods. Akin to the concept of technology styles discussed previously, the top-down methods involve carving out or adding a small number of molecules to a surface, while the bottom-up methods assemble atoms or molecules into nanostructures.

A top-down method that has been used in the electronics industry is photolithography. Photolithography is the process that transfers the geometric shape on a mask to the surface of a silicon wafer by exposure to UV light through lenses. The computer industry uses this technology to fabricate microprocessor chips. However, the use of photolithography to fabricate nanostructures is limited by the wavelength of the UV light. One modification can be made by using electron-beam lithography, which is a technique for creating fine patterns on a thin polymer film with a beam of electrons. Because electron-beam lithography is very expensive and slow, the development of soft lithography, which is a process that creates elastic stamp in order to transfer structures to a surface, allows researchers to reproduce patterns inexpensively in a wide range of materials. Nevertheless, this technique is not yet ideal for manufacturing complex multi-layered structure electronic devices. The need for methods to fabricate complex nanostructures that are simpler and less expensive has stimulated researchers to explore unconventional approaches.

Another top-down method involves using the scanning probe microscopes that were used to manipulate individual atoms to spell IBM. Researchers can manipulate atoms with an STM in three modes: pushing, pulling and sliding. Apart from mechanical manipulation, the STM can be used to assist in fabrication by chemistry catalyzing. In 1995, W. T. Muller et al proposed a method to use scanning probe microscopes in nanofabrication. They used a platinum-coated AFM tip to scan over the surface coated with a monolayer of azide (-N3) compounds. As a result, amino groups are formed by catalytic conversions of azide and can be used to generate more complex structures. Another nanofabrication method using scanning probe devices was introduced by E. S. Snow and P. M. Campbell. Their technique was to add a bias current to the AFM tip and monitor the electrical resistance of the structure during the fabrication process. When the target resistance is reached, the bias is switched off. This innovative feedback mechanism has been modified and used in later research. Recently, F. Rosei et al proposed a novel nanofabrication method for metal structures. This method uses organic molecules as templates for the rearrangement of copper atoms on a surface. At low temperature where the copper atoms are static, the template molecules can be moved away without damaging the copper surface by precisely controlling the STM tip. For more information on using the scanning probe devices in fabrication, a review by S. W. Hla and K. H. Rieder is recommended.

In contrast, bottom-up methods are truly representing a new style of technology. Although the advancement of the bottom-up methods may not yet be suitable for the production of electronic devices or allow us to replace conventional top-up methods in fabrication, researchers can inexpensively assemble atoms and molecules into nanostructures with dimensions between 2 and 10 nanometres by self-assembly chemical reactions. One innovation created with a bottom-up method is a carbon nanotube.

Iijima of NEC in 1991. A carbon nanotube is a tube-shaped carbon material that is measured in nanometre scales. It became the fifth type of solid-state carbon after diamond structures, graphite structures, non-crystalline structures and fullerene molecules or buckyballs, which were discovered by R. F. Curl et al in 1985. Since then, researchers have been studying the properties and characteristics of carbon nanotubes. Different structures of carbon nanotubes varying in length, thickness, type of spiral and number of layers have been developed for various purposes. Recently, NEC announced the world's first compact fuel cell for mobile devices that uses spiral-shaped carbon nanotubes or nanohorns for the electrodes. Carbon nanotubes are expected to be a key material in the future.

Another new material, quantum dots, is made by bottom-up methods. Quantum dots are crystals that emit only one wavelength of light when their electrons are excited. Because the electrical, magnetic and optical properties of the dot are regulated by the size of the dot, the production of quantum dots must maintain their size and composition. The size of the dots can be selected by varying the amount of time for the chemical reaction. The emission of light by quantum dots could be used in medicine as a biological marker. Alternatively, quantum dots could be used as quantum bits and to form the basis of computers.

Nanocomputers

In 1965, G. Moore, the co-founder of Intel, predicted a trend that the number of transistors contained in a microprocessor chip would double approximately every 18 months. This became known as Moore's law. As exemplified in Intel's chips, the prediction appears surprisingly correct. However, with the development of nanotechnology researchers hope to break the barrier of Moore's law.

One of the first achievements in nanocomputer research was perhaps the development of single-electron tunnelling (SET) transistors by D. Averin and K. Likharev in 198517. Later in 1987, T. A. Fulton and G. J. Dolan at Bell Laboratories fabricated single-electron transistors and made an observation on the quantum properties and effects of electrons when transistors are in operation. As techniques in nanofabrication advances, researchers have successfully created electronic components including transistors, diodes, relays and logic gates from carbon nanotubes. The next step is providing the interconnection between components. Researchers have been working on a different type of nanoscale wires called semiconductor nanowires and studied how to interconnect and integrate the components. The final step to build a computer processor is to fabricate the designed circuit. Recently, the semiconductor industry has successfully built 70-megabit memory chips containing over half billion transistors18. As the advancement in nanofabrication progresses, the silicon-based nanocomputer will step closer into reality.

Another approach to nanocomputers is DNA computing. Deoxyribonucleic acid (DNA) is a nucleic acid that carries genetic information for the biological development of life. In 1994, L. Adleman introduced the idea of solving a well-known complex mathematical problem, called the travelling salesman problem, by using DNA. His DNA computer showed that DNA could indeed be used to calculate complex mathematics; however, it is not yet comparable to conventional computer in terms of speed and ease of use. Nevertheless, his work has encouraged the development in DNA computing. In 1997, researchers at the University of Rochester built DNA logic gates, another step towards a DNA computer. The fact that a DNA molecule can store more information than any conventional memory chip and that DNA can be used to perform parallel computations make the area very appealing. Regardless of the success of DNA computers, the development of silicon-based nanocomputers could use the advantages of DNA computing.

Apart from silicon-based nanocomputers and DNA computers, researchers believe that quantum computers may be another promising approach that overcomes the limits of conventional computers. Feynman began one of the first research groups to explore computational devices based on quantum mechanics. In 1982, he demonstrated how computations could be done by quantum systems according to the principles of quantum physics. In quantum computers, the binary data in conventional computers are represented by quantum bits, or qubits, which can be in a state of 0, 1 and superposition (simultaneously both 0 and 1). As a quantum computer can hold multiple states simultaneously, it is argued that it has the potential to perform a million computations at the same time. However, quantum computers are based on quantummechanical phenomena, which are vulnerable to the effects of noise. A scheme for quantum error correction is required. Researchers have been working to overcome this obstacle. To date, quantum computing is still in the very early stages.

Nanorobots

One vision of a nanoassembler or nanorobot is a device with robotic arms, motors, sensors and computer to control the behaviour, all at the scale of nanometres. In 1992, the book called "Nanosystem" by Drexler gives an analysis of the feasibility of machine components for such nanorobots. However, even to build a molecular motor, researchers have to consider laws of thermodynamics when motors are actually in operation. Just building a miniature version of an ordinary motor is not adequate.

Recently, a controversy arose surrounding Feynman's vision of nanorobots. In 2003, an open debate through letters between K. E. Drexler and R. E. Smalley (who was awarded a Nobel Prize for the discovery of fullerenes) was presented to public. Smalley was not convinced that such molecular assemblers envisioned by Drexler are physically possible, while Drexler insists on his previous findings. Certainly, the study of similarly-sized biological machines - organic cells - suggests there may be more effective alternatives to Drexler's nanorobots. Even if nanorobots can be realised, they will not be available in the near future.

Nanomedicine

Nanotechnology promises a great future for medical research including improved medical sensors for diagnostics, augmentation of the immune system with medical nanomachines, rebuilding tissues, and tackling aging. Proponents claim that the application of nanotechonology to medicine, so-called nanomedicine, offers ultimate benefits for human life and society by eliminating all common diseases and all medical suffering. Eventually, it is argued that nanomedicine would allow the extension of human capabilities. In 2003, R.A. Freitas Jr. commented that nanometre-scale structures and devices held great promises for the advancement of medicine including advanced biosensors, smart drugs and immunoisolation therapies.

In this initial stage of nanomedicine, nanostructured materials are being tested in various potential areas; for example, tagging nanoparticles using quantum dot nanocrystals as biological markers and smart drugs that become active only in specific circumstances. In addition, researchers have found a method to control the size of densely packed DNA structures, one of nature's efficient ways for transporting gene information. This could improve the efficiency of gene therapy for medical treatment and disease prevention. It is hoped by many that the next stage of nanomedicine, where nanorobots or nanocomputers are fully available, would expand enormously the effectiveness, comfort and speed of future medicine treatments with fewer risks and costs.

BENEFITS OF COMPUTER SCIENCE FOR NANOTECHNOLOGY

Recently, M. C. Roco of the National Nanotechnology Initiative (NNI), an organisation officially founded in 2001 to initiate the coordination among agencies of nanometre-scale science and technology in the USA, gave a timeline for nanotechnology to reach commercialisation. For the next twenty years, the NNI has divided the development of nanotechnology into four generations.

The first generation, which just ended in 2004, involved the development of passive nanostructures such as coatings, nanoparticles, nanostructured metals, polymers and ceramics. At the time of writing, we are at the end of second generation, during which we have manufacture active nanostructures including transistors, amplifiers, targeted drugs, actuators and adaptive structures. Later, from the year 2010, nanotechnology should enter the third generation. It is estimated that system of nanosystems, for example: guided molecular assembling systems, 3D networking and new system architectures for nanosystems, robotics and supramolecular devices, would be developed. Finally, from the year 2020, the fourth generation of nanotechnology should be the generation of molecular nanosystems, which would integrate evolutionary systems to design molecules as devices or components at atomic levels.

To date, nanotechnology has been developed mostly from the basis in physics, chemistry, material science and biology. As nanotechnology is a truly multi-disciplinary field, the cooperation between researchers in all related areas is crucial to the success of nanotechnology. Until now, computer science has taken a role mostly in research tools, for example: a virtual-reality system coupled to scanning probe devices in nanomanipulator project. However, according to M. C. Roco, the third and fourth generation of nanotechnology would rely heavily on research in computer science.

Perhaps reflecting the extensive use of computers in the modern world, computer science is today a broad field, with many aspects that may affect nanotechnology. Earlier sections have outlined the use of graphics and imaging with nanomanipulators. Other current uses of computer science for nanotechnology include developing software systems for design and simulation. A research group at NASA has been developing a software system, called NanoDesign, for investigating fullerene nanotechnology and designing molecular machines. The software architecture of NanoDesign is designed to support and enable their group to develop complex simulated molecular machines.

However, here we focus on intelligent systems. Research in intelligent systems involves the understanding and development of intelligent computing techniques as well as the application of these techniques for realworld tasks, often including problems in other research areas. The techniques in intelligent systems comprise methods or algorithms in artificial intelligence (AI) including knowledge representation/reasoning, machine learning and natural computing or soft computing.

An exciting new development at the time of writing is a project called PACE (programmable artificial cell evolution). This large interdisciplinary project aims to create a "nanoscale artificial protocell able to selfreplicate and evolve under controlled conditions". The protocells in this work are intended to be the "simplest technically feasible elementary living units (artificial cells much simpler than current cells)".

These are intended to act as nanorobots, comprising an outer membrane, a metabolism, and peptide-DNA to encode information. Evolutionary modelling is being used extensively in PACE, to analyse real and simulated protocell dynamics, their possible evolution, and the evolution of (potentially noisy) protocellular networks. In addition to this work, computer modelling of embryogenesis and developmental systems is becoming increasingly popular in computer science. Should artificial cells become a reality, such models will provide a method for their genes to be programmed in order to enable the growth of larger, multicellular forms.

Apart from genetic algorithms and other evolutionary algorithms that have promising potential for a variety of problems (including automatic system design for molecular nanotechnology), another emerging technique is swarm intelligence, which is inspired by the collective intelligence in social animals such as birds, ants, fish and termites. These social animals require no leader. Their collective behaviours emerge from interactions among individuals, in a process known as self-organisation. This collective intelligence in social animals often cannot emerge from direct interaction among individuals. Instead, indirect social interaction (stigmergy) must be employed. Each individual may not be intelligent, but together they perform complex collaborative behaviours. Typical uses of swarm intelligence are to assist the study of human social behaviour by observing other social animals and to solve various optimisation problems. There are three main types of swarm intelligence techniques: models of bird flocking, the ant colony optimisation (ACO) algorithm, and the particle swarm optimisation (PSO) algorithm. Different techniques are suitable for different problems.

Although still a young field of computer science, swarm intelligence is becoming established as a significant method for parallel processing and simultaneous control of many simple agents or particles in order to produce a desired emergent outcome. For example, researchers at the Santa Fe Institute developed a multi-agent software platform, called Swarm inspired by collaborative intelligence in social insects, for simulating complex adaptive systems.

Likewise, BT's Future Technologies Group developed a software platform known as EOS, for Evolutionary Algorithms (EAs) and ecosystem simulations. The group uses EOS for research into novel EAs and ecosystem models and for rapid development of telecommunication related applications. Systems such as these will become increasingly important for modeling molecular machine systems. They are also being investigated as a solution to provide self-healing, adaptive and autonomous telecommunications networks. Another potential benefit of such techniques for complex adaptive systems in this area would be to control intelligently the manufacture of nanometre-scale devices, where no exact mathematical model of the system exists. Many intelligent systems' techniques have been successfully applied in control system of various complex applications. Although at nanometrescale the principles and properties of materials are altered, researchers have attempted to solve other dynamic problems using soft computing techniques and have been developing new techniques to cope with such problems.

Also inspired by emergent collaborating behaviours of social insects, the Autonomous Nanotechnology Swarm (ANTS) architecture for space exploration by NASA Goddard Space Flight Centre is claimed to be a revolutionary mission architecture. The ANTS architecture distributes autonomous units into swarms and organises them in hierarchy by using the concept of artificial intelligence. Researchers at the centre have been developing a framework to realise the autonomous intelligent system by using an Evolvable Neural Interface (ENI). As a result, the interface allows cooperation between higher-level neural system (HLNS) for elementary purpose actions and lower-level neural system (LLNS) for problem solving as required in real-world situations. In the plan, each autonomous unit will be capable of adapting itself for its mission, and the ANTS structures will be based on carbon nanotube components.

In 1996, O. Holland and C. Melhuish investigated the abilities of single and multiple agents on a task with agents under similar circumstances as future nanorobots (minimal sensing, mobility, computation and environment). The task to be solved by the agents in their studies was to learn to move towards a light source by using simple rule-based algorithms. In the case of single agents, the result was efficient, but performance degraded as the amount of noise increased. In the case of multiple agents, the best result was from the algorithm that formed collective behaviours akin to genuine social insects. This investigation showed that emergent collective intelligence from social interactions among agents modelled on social insects could cope with the limited capabilities that would be inevitable in future nanoscale robots.

Recently, B. Kaewkamnerdpong and P. J. Bentley proposed a new swarm algorithm, called the Perceptive Particle Swarm Optimisation (PPSO) algorithm. The PPSO algorithm is an extension of the conventional PSO algorithm for applications in the physical world. By taking into account both the social interaction among particles, and environmental interaction, the PPSO algorithm simulates the emerging collective intelligence of social insects more closely than the conventional PSO algorithm; hence, the PPSO algorithm would be more appropriate for real-world physical control problems. This is the first particle swarm algorithm to be explicitly designed with nanotechnology in mind. Because each particle in the PPSO algorithm is highly simplified (each able to detect, influence or impact local neighbours in limited ways) and the algorithm is designed for working with a large number of particles, this algorithm would be truly suitable for programming or controlling the agents of nanotechnology (whether nanorobots, nanocomputers or DNA computers), whose abilities are limited, to perform effectively their tasks as envisioned.

This is seen as a crucial "missing link" in bottom-up nanotechnology: the control of the nano-sized agents. A billion (or trillion) tiny particles, whether complex molecules or miniature machines, must all cooperate and collaborate in order to produce the desired end result. None will have, individually, sufficient computing power to enable complex programming. Like the growth of crystals, the development of embryos, or the intelligent behaviour of ants, bottom-up nanotechnology must be achieved through collective, emergent behaviours, arising through simple interactions amongst itself and its environment. Computer science, and especially fields of research such as swarm intelligence, will be critical for the future of bottom-up nanotech.

CONCLUSION

As the development of nanotechnology progresses in several disciplines including physics, chemistry, biology and material science, computer scientists must be aware of their roles and brace themselves for the greater advancement of nanotechnology in the future. This paper has outlined the development of nanotechnology. It is hoped that this gentle review will benefit computer scientists who are keen to contribute their works to the field of nanotechnology. We also suggested the possible opportunities that computer science can offer, which can benefit other nanotechnologists from other fields by helping them be aware of the opportunities from computer science. This paper is intended to promote collaboration between computer scientists and other nanotechnologists.

As computer scientists who are interested in the field of nanotechnology, one of our future works is to build a system that consists of a large number of particles automatically forming into a designed structure. By using the PPSO algorithm to control the swarm of particles, each particle performs lightweight computations and holds only a few values. It is anticipated that models such as these will lead to successful bottom-up nanotechnology systems in the future.

Chapter 12

Quantum Computing with Quantum Dots

INTRODUCTION

In the last years we observe a rapid progress in the theory of quantum computing. Moreover, various physical realizations of quantum computations are intensively studied. The theory of quantum computation is based on a direct application of the laws of quantum mechanics to the computations. According to our present knowledge, quantum mechanics and quantum electrodynamics provide a complete description of the structure and properties of the world at the microscopic scale, i.e., the objects of sizes smaller or comparable with sizes of molecules. In particular, the quantum theory of atoms and elementary particles (electron, photon, etc.) describes with a high precision the properties, which are verified in experiments.

Quantum computational algorithms consist of sequences of logic operations, which are performed on quantum states and implement the quantized (discrete) values of basic physical quantities, like energy and angular momentum. The quantum information is recorded with the help of quantum bits, called qubits, which are the quantum states in the two-dimensional Hilbert space. The physical realization of the qubit can be done with the use of any two-level quantum system, i.e., the microobject characterized by the observable with two discrete eigenvalues. For example, the qubit can be realized on the two spin states of the electron or the two states of the polarization of the photon. The qubits serve to storage quantum information. They can be transformed with the use of quantum logic operations. In the mathematical language, the quantum logic operations (gates) are described by the unitary transformations between the quantum states.

Parallel to the progress in the theory of quantum computation the experimental studies are recently carried out in order to find physical realizations of qubits and to perform controlled transformations on them. Several different physical systems are investigated in order to develop the realizable quantum computing technology. The interesting results have been obtained for single ions in ion traps, atoms and photons in quantum-electrodynamics (QED) cavities, molecular systems in nuclear magnetic resonance (NMR) apparatuses, and Cooper pairs in superconductors. Particularly promising is the application of semiconductor nanostructures, especially quantum dots, as the quantum computing devices. The semiconductor devices possess the advantage that the technology of their fabrication (nanotechnology) is a natural extension of the technologies used in the present computer industry and moreover they can be easily integrated with the existing hardware.

The quantum dots are the laboratory produced solid-state structures with nanometer sizes, in which the motion of charge carriers (electrons and holes) is limited in all three spatial dimensions. These are the smallest structures among the artificially fabricated objects. Their electronic properties can be modified and controlled by the modern electronic devices. We note that the quantum dots determine the limit of the current trend of miniaturization of electronic devices. This trend relies on the man-made producing of the devices with decreasing size. The smaller systems than quantum dots that can be used in future electronics (molecular electronics) are natural atoms and molecules. The quantum dots are called artificial atoms, since the confined electrons (holes) form localized quantum states with the properties similar to those of natural atoms. In particular, the energy levels associated with the quantum confined states are discrete. Applying an external electromagnetic field, we can change - in a broad range the electronic properties of quantum dots. Therefore, the quantum dots are the nanostructures, which can be used as elements of future quantum computers.

The quantum computers are nowadays at the stage of the laboratory research. A different situation takes place for nanocomputers that are nowadays introduced into the production. The basic elements of nanocomputers, i.e., a nano-field-effect-transistor (NFET) and nano-integrated circuit (nanoIC), recently reach the size below 100 nm. Therefore, quantum phenomena appear in their operation. On the contrary to quantum computers, whose operation just exploits quantum effects, the quantum effects in nanocomputers play a damaging role and limit their computational efficiency. For example, the tunnel currents spoil the isolating properties of blocking layers. The operation of nanocomputers is still based on the laws of classical physics. We also note that some possible physical realizations of quantum computers, e.g., ion traps, QED cavities, and NMR systems, are of centimeter size, which suggests that these technologies will not necessarily lead to a further miniaturization of the future computing machines. However, it is expected that any technology of quantum computation should lead to an enormous increase of the computational power.

The history of the quantum computation begins with the articles of Feynman, who as the first proposed a direct application of the laws of quantum mechanics to a realization of computational algorithms. We underly that the operation of up-to-date built computers, and also nanocomputers, can be completely described by the laws of classical physics, in particular the Maxwell equations, which are the fundamental equations of the classical electrodynamics. In spite of the fact that the present computers contain transistors, the operation of which is based on the electron band structure properties of semiconductors, the computations in the conventional computers run according to the equations of classical physics. For example, in conventional transistors the read/write operation of the single classical bit requires a flow of 10^6 ◎ 10^9

electrons. For comparison, the read/write operation of the quantum bit can be performed with the single electron.

The fundamental ideas of quantum computing were introduced and developed in the papers. A model for quantum computations and a description of the universal quantum computer as a quantum Turing machine were elaborated by Deutsch. Shor introduced the quantum algorithm for the integer-number factorization. Grover proposed the fast quantum search algorithm. Wooters and Zurek proved the non-cloning theorem, which puts definite limits on the quantum computations. Calderbank and Shor elaborated the quantum error-correcting method. Recently, the theory of quantum computing is an advanced theory, which links the elements of physics, mathematics, and computer science.

In the present article, we discuss a possibility of application of quantum dots to quantum computations. The article is organized as follows: in Section we provide a brief introduction to the quantum computing, in we present basic properties of quantum dots, in Section 4 we discuss a possible realization of qubits and quantum logic gates on two-electron spin states in coupled quantum dots, and in Section 5 we give conclusions and summary.

QUANTUM BITS AND QUANTUM LOGIC GATES

Qubits

The classical information is stored with bits, i.e., the states of the classical system, which take on two values 0 or 1, each of which occurs with probability 0 or 1. Quantum bits (qubits) are the quantum-mechanical counterparts of classical bits. The qubit is defined as a quantum state vector in the twodimensional Hilbert space H^2. If vectors $|0\rangle$ and $|1\rangle$ form the orthonormal complete basis in H^2, then the qubit can be written down as

$$|\psi\rangle = c_0 |0\rangle + c_1 |1\rangle,$$

where the complex probability amplitudes c_0 and c_1 satisfy the normalization condition $|c_0|^2 + |c_1|^2 = 1$. The set of states $\{|0\rangle, |1\rangle\}$ is called a computational basis.

The information capacity of classical and quantum bits is different. Contrary to the classical bit, which can be in classical state 0 or 1 with probability 1, the quantum bit takes on a continuum of values, which are determined by the amplitudes c_0 and c_1. However, these amplitudes are non-measurable. If we perform a measurement on qubit (1), we obtain either outcome 0 with probability $|c_0|^2$ or outcome 1 with probability $|c_1|^2$.

However, if the quantum system is described by the qubit being exactly equal to one of the states of the computational basis, i.e., $|\psi\rangle = |0\rangle$ or $|\psi\rangle = |1\rangle$, then we can predict the exact result of the measurement with probability 1. This dichotomy between the non-observable general state of the qubit and the precise result of the measurement in the basis state (eigenstate of the observable) plays an essential role in quantum computations.

For the quantum computing, besides the single qubit states (1), we also need two-qubit states, which are the states of the two-particle quantum system. The two-qubit states can be constructed as tensor products of basis states $\{|0\rangle, |1\rangle\}$. Accordingly, the two-qubit basis consists of the states $|00\rangle$, $|01\rangle$, $|10\rangle$, $|11\rangle$, where we apply a shortened notation, e.g., $|00\rangle \equiv |0\rangle|0\rangle \equiv |0\rangle \otimes |0\rangle$. The arbitrary two-qubit state has the form

$$|\Psi\rangle = c_0 \, |00\rangle + c_1 \, |01\rangle + c_2 \, |10\rangle + c_3 \, |11\rangle,$$

where the normalization condition takes on the form

$$|c_0|^2 + |c_1|^2 + |c_2|^2 + |c_3|^2 = 1.$$

Spin qubits

A particle with non-zero spin is particularly suitable for the physical realization of the qubit. The qubits can be formed from the spin states of the single electron, single nucleus, pair of electrons, or electron-hole system (exciton). In the present paper, we consider the particle with spin quantum number 1/2, e.g., the electron, for which the z component of the spin takes on the two values $\pm (\hbar/2)$. The operator of the z spin component is $s_z = \frac{\hbar}{2}\sigma_z$,

where σ_z is the z Pauli matrix

$$\sigma_z = \begin{pmatrix} 1 & 0 \\ 0 & -1 \end{pmatrix}$$

The corresponding eigenequations have the forms

$$s_z\,|0\rangle = +\frac{\hbar}{2}|0\rangle,\; s_z\,|1\rangle = -\frac{\hbar}{2}|1\rangle.$$

The eigenstates can be written in the form of spinors, i.e.,

$$|0\rangle = \begin{pmatrix} 1 \\ 0 \end{pmatrix},\; |1\rangle = \begin{pmatrix} 0 \\ 1 \end{pmatrix}.$$

Another physical quantity of interest is the spin magnetic dipol, which possesses the z component

$$\mu_z = -\frac{1}{2} g^* \mu_B \sigma_z,$$

where μ_B is the Bohr magneton ($\mu_B = 0.927 \times 10^{-23}$ Am^2), g^* is the effective Lande factor, which in semiconducting materials can take on positive as well as negative values, e.g., for the electron in Si $g^* = 1.998$, in Ge $g^* = 1.563$, and in GaAs g^* = "0.44. For comparison, for the electron in the vacuum g^* = 2.0.

The spin can be experimentally detected using the interaction of the spin magnetic dipol with the external magnetic field B. For B = (0, 0, B) the Hamiltonian of this interaction has the form

$$H_{int} = -\mu_z B = \frac{1}{2} g^* \mu_B \sigma_z B\,.$$

If the quantum system possesses energy E_ν in the absence of the external magnetic field, then - according to (3), (5), and (8) - the interaction of the spin magnetic dipol with the magnetic field leads to the splitting of this energy level into the two spin sublevels with energies

$$E_{\nu\pm} = E_\nu \pm \frac{1}{2} g^* \mu_B B,$$

where sign + corresponds to state $|0\rangle$ with spin $+\hbar/2$ and sign – corresponds to state $|1\rangle$ with spin $-\hbar/2$. Eq. (9) describes the spin Zeeman effect, which can be observed by the spectroscopic methods. For example, for Si at B = 10 T the spin splitting energy is ~0.6 meV, which corresponds to the radiation with the wave length ~2 mm.

Quantum Logic Gates

The qubits can be transformed using the quantum logic gates, which are performed with the help of unitary transformations U, which transform initial state $|\psi_i\rangle$ into final state $|\psi_f\rangle$ according to

$$|\psi_f\rangle = U|\psi_i\rangle.$$

Depending on the type of qubit, on which they operate, we deal with either one- or two-qubit gates. The quantum NOT gate, being a counterpart of the classical NOT gate, defined as

$$U_{NOT} \equiv \begin{pmatrix} 0 & 1 \\ 1 & 0 \end{pmatrix},$$

is an example of the one-qubit gate. If we write the one-qubit state (1) in a matrix form as

$$|\psi\rangle = \begin{pmatrix} c_0 \\ c_1 \end{pmatrix},$$

then the NOT gate operates on the one-qubit state as follows:

$$U_{NOT}\begin{pmatrix} c_0 \\ c_1 \end{pmatrix} = \begin{pmatrix} c_1 \\ c_0 \end{pmatrix}$$

As a result, the basis states $\{|0\rangle, |1\rangle\}$ have been interchanged, i.e., $|0\rangle \leftrightarrow |1\rangle$.

The two-qubit gate operates on the two-qubit state $|\beta_1, \beta_2\rangle \equiv |\beta_1\rangle|\beta_2\rangle$, where $\beta_1, \beta_2 = 0, 1$. An example of the important two-qubit gate is the controlled-NOT gate U_{CNOT}, for which the first qubit ($|\beta_1\rangle$) is the control qubit and the second qubit ($|\beta_2\rangle$) is the target qubit. The controlled-NOT gate transforms the two-qubit basis states as follows:

$U_{CNOT}|00\rangle = |00\rangle$, $U_{CNOT}|01\rangle = |01\rangle$,

$U_{CNOT}|10\rangle = |11\rangle$, $U_{CNOT}|11\rangle = |10\rangle$,

which means that the CNOT gate changes the second qubit if and only if the first qubit is in state $|1\rangle$.

It was shown that the set of logic operations, which consists of all the one-qubit gates and the single two-qubit gate U_{CNOT} is universal in the sense that all unitary transformations on N -qubit states, where N is arbitrary, can be expressed with the help of different compositions of the gates, which belong to the universal set of gates. Another important property of quantum computations is a quantum paralelism, which is based on the fact that the single unitary transformation can simultaneously operate on all the qubits in the system. The paralelism of quantum computations is an immanent characteristic of the quantum system; therefore, no special technology is necessary for its implementation.

Conditions for the physical realization of quantum computing

The quantum computing technology has to satisfy the following conditions: (i) the physical realizability of the qubits; (ii) the possibility of the precise preparation of the initial qubit state; (iii) the controlled unitary evolution of the qubit; (iv) the possibility of the accurate measurement of the final qubit state. When considering a physical system chosen to perform the quantum computations, we have to define the physical properties, which we will exploit in computations, e.g., spin of the electron, atomic nucleus, or photon, and the physical processes for the read/write operations, e.g., radiative transitions with the emission/absorption of photons.

The conditions listed above put certain limitations on the quantum computing technology. When designing the physical apparatus, which will per form the quantum computations, we face the problem of maintaining the con trolled unitary evolution of the quantum system until the computations are completed. Such controlled evolution is possible provided that the quantum system is completely isolated from the environment. However, the complete isolation of the quantum-computing system disables the read/write operations. Therefore, some slight interaction of the quantum system with the environment is necessary. On the other hand, this interaction leads to decay and decoherence processes, which reduce the performance of the quantum computer. In the decay process, the quantum system goes over - in a very short time - to a new state releasing a part of its energy to the environment. For example, the change of spin state $|0\rangle \rightarrow |1\rangle$ is accompanied by the emission of the photon. The decay is characterized by the decay time (relaxation time), which for the spin states can be very long. The recent measurements of the Zeeman splitted spin states in quantum dots give a lower bound of 50 μs on the relaxation time at $B = 7.5$ T.

A decoherence is the much subtler effect, in which the energy is conserved but the relative phase of the different basis states of the qubit is changed. As a result of decoherence the qubit changes as follows:

$$|\psi\rangle \rightarrow c_0|0\rangle + e^{i\theta} c_1|1\rangle,$$

where the real number θ denotes the relative phase. The appearance of the non-zero relative phase results from the coupling of the quantum system with the environment and can lead to essential changes in the measurement statistics. For example, the quantum-mechanical expectation value of the measured quantity is changed. The decoherence time t_{decoh} is usually much shorter than the decay time; therefore, the decoherence can be treated as the most detrimental effect for the quantum computations. The ratio of the decoherence time t_{decoh} to the elementary operation time t_{oper}, i.e., $R = t^{decoh,}\, t_{oper}$

$$R = \frac{t_{decoh}}{t_{oper}}$$

is an approximate measure of the number of computation steps performed before the coupling with the environment destroys the qubit. For different quantum-computing technologies this ratio changes in broad limits :

$10^3 \leq R \leq 10^{13}$. For example, $R = 10^3$ for the electron states in quantum dots, $R = 10^7$ for nuclear spin states, and $R = 10^{13}$ for trapped ions.

QUANTUM DOTS

A semiconductor quantum dot is the nanostructure, the linear size of which does not exceed 1 μm in each spatial direction. The typical sizes of the quantum dots are between ~10 nm and ~100 nm. The potential created in the quantum-dot nanodevice limits the charge carrier motion in all the three dimensions. This confinement potential possesses the range comparable with the size of the quantum dot and the finite depth. The typical depth of the confinement potential, i.e., the electron potential energy minimum measured with respect to the conduction band bottom of the embedding material, is of the order of ~0.1 eV to ~1 eV. This leads to the energy separations between the one-electron energy levels of the order of few meV.

These energy separations put an additional limitation on the realizability of quantum computations, namely, in order to avoid thermal excitations, we have to maintain the temperature of the nanodevice below 1 K. There are many types of quantum dots, among which, the best candidates for the possible implementation of quantum logic gates are the electrostatic (gate controlled) quantum dots. The electrostatic quantum dot consists of the sequence of vertically stacked layers, which form single or multiple potential wells and barriers.

The source and drain electrodes are located at the bottom and top sides of the layer sequence. The entire quantum-dot nan- odevice usually possesses a cylindrical symmetry and can have either a form of an etched pillar or a layer sequence with a metal cap. Depending on the number of barrier layers, the nanodevice can contain either a single or multiple quantum dot. In the pillar-shape quantum-dot nanodevice, an additional gate electrode is placed at the cylinder side, which increases the ability of tuning of the electrostatic field in the quantum dot. In the electrostatic quantum dot, the confinement potential results from both the conduction band offsets and the external electrostatic field created by the electrodes.

The knowledge of this potential is important for studying and modelling the electronic properties of the quantum dot. The confinement potential cannot be directly measured, but can be calculated from the first principles of electrostatics by solving the Poisson equation for the entire nanostructure. Such calculations were performed for the two different types of the electrostatic quantum dots. These were the pillar-shaped quantum dots of Tarucha et al. and capped quantum dots of Ashoori etal. The results obtained show that the confinement potential V can be parametrized by either the Gaussian function or power-exponential function of the form

$$V = -V_0 \exp[-(r/|R)^p - (|z|/Z)^p],$$

where $V_0 > 0$ is the depth of the potential well, $r = \sqrt{x^2 + y^2}$, $p > 1$, R and Z are the measures of the confinement potential range in the lateral directions x, y and vertical direction z, respectively. For $p = 2$ we obtain the Gaussian potential and for $p > 10$ the shape of the confinement potential resembles the rectangular potential well.

Electrons confined in the quantum dot form localized bound states with discrete energy levels. These states exhibit a qualitative similarity to the quantum states of natural atoms. Therefore, the quantum dots are some-times called artificial atoms. The two quantum dots, which are coupled by the tunnel barrier, form an artificial molecule. From the point of view of a possible application to quantum computation, the single-electron trans- port via the quantum dot is of crucial importance.

The main single-electron transport channel is the sequential tunneling, in which the single electrons tunnel through the dot in subsequent time intervals provided the transport conditions are fulfilled. The single-electron transport measurements appeared to be the successful spectroscopic method, which allowed to discover the wonderful properties of quantum dots: the filling of the shells of artificial atoms and the quantum Coulomb blockade. The quantitative theo-retical description of these effects was given in paper. The vertical gated quantum-dot nanodevice is a prototype of a single-electron transistor,which can be switched on and off by the flow of the single electron. In future, the application of single-electron transistors will lead to a much higher performance of electronic devices at greatly reduced power consumption.

Recently, the possibility of implementation of quantum

dots to quantum computation is intensively studied. The qubits can be realized as either the charge states or spin states of the quantum dots. The electro-static quantum dots seem to be especially well suited to perform the quantum computations, since their electronic properties can be modelled by the proper choice of the nanostructure parameters and tuned by changing the external voltages applied to the electrodes. This enables both to obtain the designed properties of the quantum states (quantum engineering) and perform the controlled logic operations on these states. Moreover, the modern nanotechnology of fabrication of quantum dots is an extension toward a smaller feature size of the well known semiconductor MOSFET technology. Therefore, its introduction into the production is more easy than those of the other quantum-computing technologies, based on ion traps and QED cavities, which are obtained only in advanced laboratories.

MODEL OF IMPLEMENTATION OF QUBITS AND OGIC GATES WITH QUANTUM DOTS

In quantum dots, the qubits can be formed from the quantum states of electrons or excitons. The exciton is the bound electron-hole system, which is created in the absorption of the photon with energy comparable with the forbidden energy gap of the semiconductor. After the lifetime of the order of microseconds, the exciton recombinates with the emission of the photon. The excitonic qubit can be realized in a simple manner: we ascribe state $|1\rangle$ to the existing exciton in the ground state and state $|0\rangle$ to the system after recombination, i.e., an empty conduction band, a fully filled valence band, and a photon. This concept has a disadvantage of the short decay time (lifetime) of the exciton. However, it has the important advantage of the easy performance of the read/write operations with the help of visible light photons in emission/absorption processes. A possible implementation of biexcitons (bound two-exciton systems) as the qubits has also been studied.

Nowadays, it seems that the most promising for quantum

computation is the application of the spin states of quantum-dot confined electrons.

The electron spin states possess the following advantages: very long relaxation time in the absence of external fields, fairly long decoherence time (the experiments suggest $t_{decoh} \simeq 1\mu s$), and the possibility of easy manipulation of the spin by the external magnetic field. The research toward the implementation of the electron spin as a new information carrier is the subject of a new electronics based on spin, called spintronics.

Recently, the spin transistor has been designed. We can use the electron spin states to construct qubits and logic operations on them in two ways, either directly with the application of the spin magnetic dipol coupling with the magnetic field or indirectly with the application of symmetry properties of the many-electron wave function. The indirect application of spin to represent the qubit is based on the change of the sign of the wave function during the exchange of the space-spin coordinates of two electrons. This fundamental quantum-statistical property leads to the well defined symmetry of the spin states in a subspace of states with the definite spin.

For example, in the two-electron system, the spin singlet state (antisymmetric against the spin exchange) possesses a different (usually lower) energy than the spin triplet states (symmetric with respect to the spin exchange). The resulting singlet-triplet energy level splitting can be used to distinguish the spin qubits and to perform logic operations on them, e.g., with the use of photon emission and absorption.

In the present section, we discuss a modification of the proposition of an implementation of spin states of electrons confined in coupled quantum dots in a quantum computation. In this case, the computational basis is formed from the eigenstates of the z component of the electron spin. Due to the universal character of the controlled NOT gate we will discuss its realization with the use of the two coupled quantum dots. The operation of the CNOT gate is defined by Eq..

Let us consider the two-electron system in the two coupled

quantum dots labelled by index $j = 1, 2$. We assume that the single electron is confined in each dot. If the quantum dots are fabricated from the different materials, the effective Lande factors are different, i.e., $g_1^* \neq g_2^*$. According to Eq. (9), this leads to the different Zeeman energy of the electron in each dot. The similar effect can be achieved by applying the inhomogeneous magnetic field, which takes on the different value in each of the dots. The material composition and the thickness of the barrier layer, which separates the quantum dots, is chosen so that a small coupling between the dots is ensured. The Hamiltonian of the system considered has the form

$$H = H_1 + H_2 + H_{int},$$

where H_j (j = 1, 2) is the one-particle Hamiltonian of the electron with spin $s_{z,j}$ in the external magnetic field, i.e.,

$$H_j = \omega_j\, s_{z,j},$$

where $\omega_j = g_j^* \mu_B B_j / \hbar$ and the magnetic field in the jth quantum dot is given by $B_j = (0, 0, B_j)$. The interaction Hamiltonian has the form

$$H_{int} = (4/\hbar)\Omega s_{z,1}\, s_{z,2},$$

where parameter Ω characterizes the coupling between the electron spins in different quantum dots. The spin operators $s_{z,1}$ fulfill the eigenequations

$$s_{z,1}|0, l\rangle = +\frac{\hbar}{2}|0,l\rangle,\ s_{z,1}|1, l\rangle = -\frac{\hbar}{2}|1,l\rangle,$$

where $|k, l\rangle$ are the two-spin states ($|k, l\rangle = |k\rangle \otimes |l\rangle$) and $k, l = 0, 1$. The similar two equations are satisfied for the operators $s_{z,2}$. According to the above assumptions $\omega_1 \neq \omega_2$.

In the absence of coupling, i.e., for $\Omega = 0$, Hamiltonian possesses the following eigenvalues: $\epsilon_1 = -(\hbar/2)(\omega_1 + \omega_2)$ in state $|1, 1\rangle$, $\epsilon_2 = (\hbar/2)(\omega_1 - \omega_2)$ in state $|1, 0\rangle$, $\epsilon_3 = (\hbar/2)(\omega_1 - \omega_2)$ in state $|0, 1\rangle$, and $\epsilon_4 = (\hbar/2)(\omega_1 + \omega_2)$ in state $|0, 0\rangle$. The coupling can be switched off by applying the barrier layer with sufficiently large thickness. Each spin state can be selectively addressed in processes of absorption or emission of photons with the proper frequency. For example, in the absorption process, the photon with frequency ω_1 operates exclusively on the first qubit changing it from state $|1, l\rangle$ (spin-down state) into state $|0, l\rangle$ (spin-up state). The reverse process, i.e., $|0, l\rangle \rightarrow |1, l\rangle$, is also possible in the stimulated emission. These processes occur for each l = 0, 1, i.e., for an arbitrary state of the second qubit. Similarly, the photon with frequency ω_2 will change the state of the second qubit only.

The case of the coupling switched on ($\Omega > 0$) can be described using the solutions to the eigenproblem of Hamiltonian. It appears that the two-spin basis states

$$\{|1, 1\rangle, |1, 0\rangle, |0, 1\rangle, |0, 0\rangle\}$$

are the eigenstates of Hamiltonian to the eigenvalues $E_1 = \epsilon_1 + \hbar\Omega$, $E_2 = \epsilon_2 - \hbar\Omega$, $E_3 = \epsilon_3 - \hbar\Omega$, $E_4 = \epsilon_4 - \hbar\Omega$, respectively. This means that after switching on the interdot coupling the energy levels are changed by $\pm\hbar\Omega$, while the eigenstates remain unchanged. This enables us to realize the selected transitions between the basis states. The resonant radiation with accurately chosen frequency causes the switching on and off between basis states associated with the change of one qubit depending on the state of the second qubit. For example, the photon with frequency $\omega_2 - 2\Omega$ induces the switching between states $|1, 0\rangle$ and $|1, 1\rangle$ only, leaving states $|0, 0\rangle$ and $|0, 1\rangle$ unchanged. Therefore, in the stimulated emission/absorption processes induced by this photon, we can implement the CNOT operation.

This proposition of the implementation of the CNOT logic gate is based on a simplified model of interaction between the spins. On the other hand, interaction Hamiltonian possesses the form of the Heisenberg Hamiltonian, which is an universal Hamiltonian of interaction between particles with spin. The Heisenberg Hamiltonian can be used to a description of the inter action between different physical objects with non-zero spin, e.g., electrons, ions, and atomic nuclei. The model of the CNOT gate, discussed above, can be realized in the coupled quantum dots as well as in the NMR system.

We note that the other methods are also studied for the implementation of spin qubits in quantum dots. They include: (i) a measurement of spin via the measurement of charge, (ii) a measurement of a spontaneous magnetisation of the quantum dot, (iii) electron spin resonance, and (iv) a measurement of singlet-triplet splitting with the help of a Faraday rotation. Method (i) employs the spin filter, which generates a current of spin polarized electrons through a semiconductor. Recently, a very interesting realization of qubits on quantum dots has been experimentally studied with the direct use of the electron charge.

The authors succeeded in forming the qubits in the coupled quantum dots as the states with the presence or absence of the single electron in the first or the second dot. This achievement is very promising since it enables us to construct the qubits by charging the quantum dot and detect them by measuring the electric charge.

CONCLUSIONS AND SUMMARY

The present stage of research toward a construction of a quantum computer can be characterized as follows: the theory of quantum computation, based on the solid foundations of quantum mechanics, is very advanced, while the practical realization of the quantum computation, although being a subject of intensive experimental studies in physical laboratories, is in an emerging stage. The quantum computing theory comprises the elements of quantum physics, mathematics, and information theory. This theory shows that it is possible to apply directly the quantum effects to a superdense information storage and massively parallel computing with a very high speed. Nowadays, we face the problem of the physical realization of the quantum computer. Besides the quantum-dot quantum computer, treated in the present article, the different competitive technologies are intensively studied. The most promising of them are ion traps, QED cavities, NMR systems, and superconducting systems.

The spin of quantum particles (electron, atomic nucleus, photon) is especially suitable to construct of and operate with the qubits. In the field of semiconductor physics and electronics, the application of electron spin as an information carrier has led to an emergence of spin electronics (spintronics). On the other hand, the miniaturization of semiconductor devices resulted in a development of nanoelectronics. In the nanoelectronic devices, the charge of single electrons is used to carry on and transform the information. The quantum dots are very important nanostructures from the point of view of both the spintronics and nanoelectronics. They possess the important advantage, namely, the possibility of tuning and controlling their electronic properties by changing the external electromagnetic fields. This allows us to obtain the required characteristics and modify them in a very short time.

It seems that - due to their flexibility - the quantum dots are promising candidates as elements of future quantum computers. The other important characteristics of quantum-dot nanodevices is their compatibility with the existing hardware. It is important for the emerging quantum technology that the construction of the spin- and nanoelectronic devices is based on the progress in the existing electronics and the present semiconductor industry.

In summary, nowadays we have at disposal the advanced quantum computation theory. However, the physical implementation of quantum compu-tation is in the stage of laboratory studies. The electron spin states in the coupled quantum dots, discussed in the present paper, are good candidates for the qubits. Moreover, the quantum logic gates can be implemented with these states.

Chapter 13

Molecular Electronic Crossbar Memory Circuits

INTRODUCTION

In chapter, described the fabrication procedures for molecular switch tunnel junction (MSTJ) device at a single device level and also proved that bistable rotaxane plays a crucial role in a decent conductance switching. The rotaxane molecular switches hold several advantages over more traditional switching components such as ferroelectric and ferromagnetic materials. However, a key application of these molecular switches is related to the extreme scaling of electronic circuitry to near molecular dimensions: since conductance switching within an MSTJ originates from the electrochemically driven molecular mechanical isomerization of the molecules, the switching relies purely on individual molecular properties. In both molecular dynamics (MD) and electrochemical investigations, the two molecular co-conformers are characterized by different HOMO-LUMO gaps, and therefore different tunneling probabilities. By contrast, the switching of solid state materials such as ferroelectrics involves altering the polarization state of the crystallographic lattice upon the application of an external electric field. This polarization disappears for domains below a certain critical size. An analogous phenomennon is the transition from ferromagnetic behaviour to superparamagnetic behaviour for magnetics, as the size of the ferromagnetic material is reduced. Therefore, these devices have a critical limitation in the scale-down. Second, solid state material switching depends upon a field-driven nucleation process and can be statistical in nature, especially for small crystallographic domain sizes. The molecular switches discussed in this thesis, by contrast, switch based upon electrochemical processes.

These are current and voltage driven, and depend upon molecular properties such as redox potentials, molecular orbital energies, etc. Therefore, switching voltages for the ferroelectric device could vary from junction to junction, or during many cycles. In particular, at a smaller dimension, the reliability issues are expected be more serious.

When the solid state materials are integrated into a 2D crossbar circuit that is a main architecture of the memory devices described here, their irregular switching characteristic will cause an additional problem, that is, 'a half select issue. In the crossbar memory circuit, a switching voltage, V_A, is split into two components, $+½\ V_A$ and $-½V_A$, which are then applied to the top and bottom electrodes that define a designated cross-junction. As all of the junctions are interconnected in a crossbar circuit, every junction is subject to at least some field. For the case of the solid state materials, the field generated by $\pm\ ½\ V_A$ is occasionally sufficient to perturb the state of the nucleation event. This half-select problem is considered as a generic problem for the field-poled devices that function within a 2D crossbar circuit.

Taking advantage of the switching molecules, the next challenge is to integrate those switching molecules and electrodes into fully functional circuits, patterned at nanometer dimensions. When such a memory circuit is combined with other nanoscale functional circuits such as logic and routing, computing at nanometer dimensions, which is currently unachievable with standard CMOS technology, could be realized. Entire circuits for nanoscale computing were proposed as conceptualized in figure. While efforts in my research group have focused on developing and integrating the various components of this nano-computer, this chapter will address fabrication and testing of crossbar memory circuits exclusively.

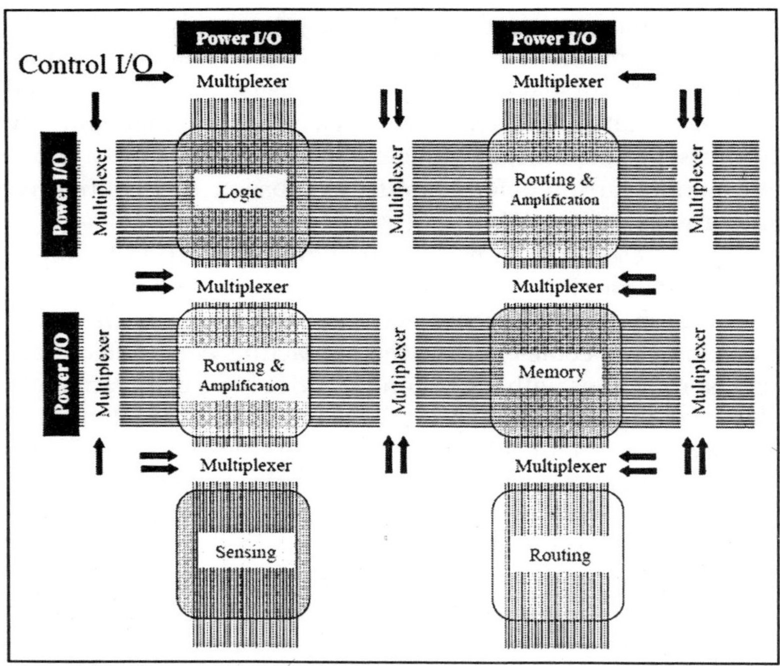

Fig. A Nanoscale Molecular Computational Platform.

Tranditional computing functions are coupled to non-traditional elements, including sensors, actuator, etc., and are illustrated as individual tiles in a mosaic-like architecture. Muliplexers and demultiplexers control communication between various functions and provide the user interfaces. Copyright 2006 Royal Society of Chemistry.

As presented in figure , the proposed computing platform is based on the crossbar architecture. The crossbar geometry provides a promising architecture for nanoelectronic circuitry. The crossbar is tolerant of manufacturing defects a trait that becomes increasingly important as devices approach macromolecular dimensions and non-traditional (and imperfect) fabrication methods are employed.

For example, Teramac had nearly a quarter million hardware defects and yet could be configured into a robust computing machine. The crossbar is a periodic array of crossed wires, similar to a two-dimensional crystal, implying that non-traditional methods can be employed for its construction. Finally, the crossbar is the highest density, two-dimensional digital circuit for which every device can be independently addressed. This attribute enables the circuit to be fully tested for manufacturing defects and to be subsequently configured into a working circuit.

For these reasons, my research group has tried to utilize the crossbar architecture for memory circuits. The progress in this crossbar memory project has been made both in the switching molecules and in the contacting electrodes: more robust rotaxanes with higher on/off ratios have been rationally designed and synthesized. At the same time, as an attempt to constitute a circuit with higher density, various electrode-patterning techniques have been developed and tested. In practice, the total number of bits fabricated within a single crossbar circuit increased from 16 to 64 to 4,500 to 160,000 bits as the main electrode pattering techniques were improved from photo-lithography to electron-beam lithography (EBL) to recently developed nanowire array technique (figure), and as methods to increase the compatibility of our nanofabrication procedures with the molecular switch components were improved.

Notably, the resolution of the nanowire array that has been developed by my research group is far beyond that of the conventional EBL. Hence, the memory circuits utilizing this nanowire array technique set a remarkable landmark in memory bit density. In this chapter, I will first describe the nanowire array technique, so called superlattice nanowire pattern transfer (SNAP) method, and then focus on fabrication and testing of 160 kbit crossbar memory circuits based on the SNAP technique and rotaxanes.

The fabrication of the 160 kbit memory circuits at a bit density of 10^{11} bits/cm^2 was totally nontrivial. At this point, reviewing the history of molecular memory projects in my research group provides guidance for understanding our efforts toward device miniaturization, as well as understanding difficulties of the fabrication procedure. Until reaching this unexplored bit density, many scientists from several groups have contributed to different components of the circuits at each stage: In conjunction with Hewlett-Packard group, my research group initially proposed the concept of the defect-tolerant crossbar architecture.

Since then, significant progresses in achieving the actual molecule-based devices were initiated by the former postdocs in my research group, Dr. Collier, Dr. Wong and Dr. Luo. They optimized conditions for many of the key fabrication steps, including monolayer deposition by the Langmuir-Blodgett technique and metal deposition to form the top electrical contact to the molecules. They also established rational electrical measurement schemes such as the remnant molecular signature scan and the temperature-dependent volatility scan to test the switching and mechanism.

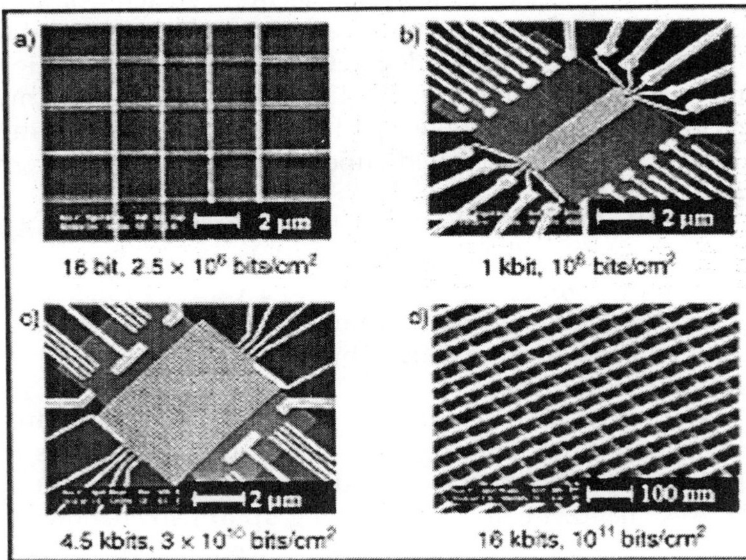

Fig.. A series of crossbar Molecular Electronic memory circuits. These circuits are arranged a-d in Accordance with the Chronology of their Fabrication. The total number of bits and the memory density of each memory circuit were denoted below scanning Electron Micrograph (SEM) Pictures.

They also demonstrated relative robust molecular switches that could be cycled > 1,000 times. Dr. Luo developed next-generation fabrication procedures for the memory circuits based on the EBL-defined electrodes. He demonstrated that the molecular switching in EBL-defined circuits is still very robust, and he developed a number of procedures for the integration of Si SNAP nanowire bottom electrode arrays with EBL defined top electrodes. For my part, I was teamed up with another group member, Jonathan Green, to achieve the large-scale (160,000 bit) SNAP nanowire-based memory circuits at extreme density (10^{11} bitscm^{-2}).

SUPERLATTICE NANOWIRE PATTERN TRANSFER (SNAP) METHOD

As described in the previous subchapter, the scaling advantage of a rotaxane molecular electronic switch would be best illustrated only when electrodes with molecular dimensions are integrated together to form the junction sandwiching the molecule. For this and other reasons, my research group has developed unique nanowire array fabrication technique called superlattice nanowire pattern transfer (SNAP).

For the SNAP method, the layer structure of a GaAl/ $Al_xGa_{(1-x)}$ As superlattice in which each layer is grown under atomic-level control, is translated into a variety of metals or silicon. The width and pitch of final nanowire array made of metals and/or silicon are defined by the initial superlattice film widths and spacings.

With the SNAP method, my research group has demonstrated the fabrication of silicon nanowire arrays in which each wire is about 8 nm wide and at a pitch of about 16 nm. Also, nanowire arrays containing up to 1,400 nanowires have been demonstrated. Moreover, in comparison to other nanowire growth methods, such as the vapour-liquid-solid growth method most fully developed in the Lieber group, the SNAP method has no limitation in length of the nanowires within arrays. A few millimeter long nanowire arrays are routinely produced.

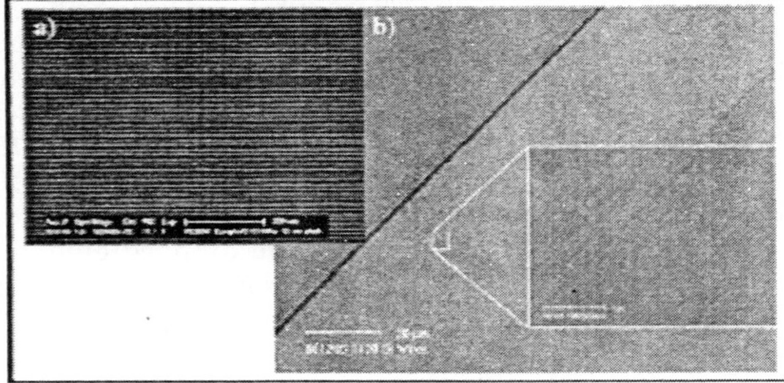

Fig.. SEM images of Si Nanowire arrays of (a) 15 nm pitch and (b) 1400 wires.

Here, I briefly go through the SNAP procedures. More detailed procedures of the SNAP method are described in the thesis of Jonathan Green, who was also involved in the project as a leading member. First, a wafer containing the superlattice was diced into small pieces, which are referred to as masters. When turned on its side, a master is largely a GaAs wafer, with the top edge of the wafer containing the superlattice. When viewed from the edge, this superlattice structure is like a club sandwich, with alternating layers of GaAs and $Al_xGa_{(1-x)}As$ films substituting for the meat and bread layers in the sandwich. It is this superlattice edge that provides the initial template for the nanowires. This edge is first cleaned carefully in a class 1000 clean room using methanol and gentle swabbing, so that it is clean by eye when viewed under an optical microscope. Once the cross-section turns out to be completely dust-free, a dilute mixture of buffered hydrofluoric acid (15 ml of 6:1 buffered oxide etchant, 50 ml of H_2O) is used to selectively etch the $Al_xGa_{(1-x)}As$, thus forming a comb-like structure as shown in figure 3-4b. Conversely, the GaAs could be selectively etched to form a complementary comb-like structure. Next, about 10 nm of Pt layer was deposited onto the superlattice side at about 45° tilted angle using electron beam evaporation of a Pt target. This angle could be varied to yield some control over the nanowire width. In preparing silicon substrates (using silicon-on-insulator wafers), the substrates were cleaned by rinsing with a series of solvents until the surface becomes completely dust-free. A mixture of epoxy and poly (methyl methacrylate) (PMMA) (10 drops of epoxy and 1 drop of curing agent, 0.18 g of 6 % (in weight) PMMA in chlorobenzene, additional 10 ml of chlorobenzene) was spincast onto prepared substrates at 8000 rpm. Pt deposited masters were dropped onto the epoxy coated substrates so that the superlattice side contacted and adhered to the epoxy layer.

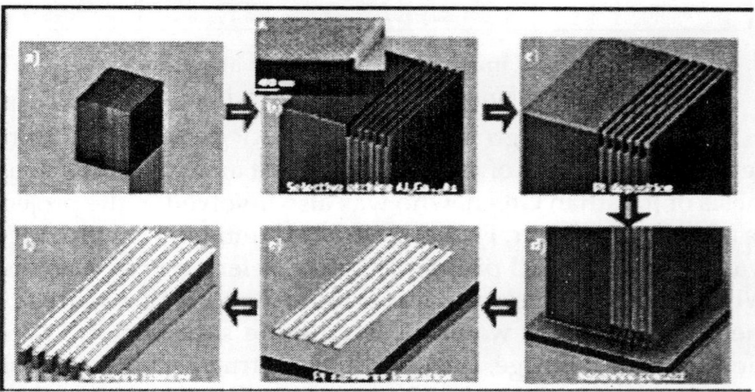

Fig. SNAP process flow. (a) The wafer Containing the Superlattice was diced into small pieces and the Superlattice side was cleaned Thoroughly. (b) $Al_xGa_{(1-x)}As$ was Selectively wet-eched. (c) A Pt layer was deposited onto the superlattice side by electron beam evaporation. (d) Masters were dropped onto the epoxy coated substrates. (e) The superlattice was removed by a wet-etch, leaving the Pt nanowirearray structure on the substrates. (f) The Pt nanowire array structure was transferred to form an array of aligned and high aspect ratio silicon nanowires via RIE.

The epoxy was cured around 130 °C for half an hour or so. The substrates were then dipped into an etching solution for removing GaAs masters, but leaving behind the Pt nanowires, which were epoxy-adhered to the substrate. After ~ 4 hr of wet-etching, the GaAs masters were peeled off and the Pt nanowire array structure remained on the substrate.

Once the quality of the Pt nanowire array was confirmed by SEM, the Pt pattern was transferred into the underlying silicon-on-insulator substrate by reactive ion etching (RIE) (CF_4:He = 20:30 sccm, 5 mTorr, 40 W, ~ 4 min for 33 nm Si layer). Upon removing Pt by aqua resia (HCl:HNO_3 = 2:1 in volume), the Si nanowire array structure was complete.

160 KBIT MOLECULAR ELECTRONIC MEMORY CIRCUITS:

The primary metric for gauging progress in the various semiconductor integrated circuit (IC) technologies is the spacing, or pitch, between the most closely spaced wires within a dynamic random access memory (DRAM) circuit. Modern DRAM circuits have 140 nanometer (nm) pitch wires and a memory cell size of 0.0408 square micrometers (ìm^2). Improving IC technology will require that these dimensions decrease over time.

However, by year 2013 a large fraction of the patterning and materials requirements for constructing IC technologies are currently classified as having 'no known solution'. Nanowires, molecular electronics, and defect tolerant architectures have been identified as materials, devices, and concepts that might assist in continuing IC advances. This belief has largely been bolstered by single device or small circuit demonstrations.

The science of extending such demonstrations to large scale, high density circuitry is largely undeveloped. In this and following sections, I describe a 160,000 bit molecular electronic memory circuit, fabricated at a density of 10^{11} bits/cm^2 (pitch = 33 nm; memory cell size = 0.0011 μm^2), which is roughly analogous to a projected year 2020 DRAM circuit. A monolayer of bistable, rotaxane molecules described in chapter 2 served as the data storage elements. Although the circuit had large numbers of defects, those defects could be readily identified through electronic testing and isolated using software coding. The working bits were then configured to form a fully functional random access memory circuit for storing and retrieving information.

A few groups have reported on non-lithographic methods for fabricating crossbar circuits, but most methods are not yet feasible for fabricating more than a handful of devices. Furthermore, the assembly of nanowires into narrow pitch crossbars without electrically shorting adjacent nanowires remains a challenge.

Despite these challenges, my research group developed the SNAP method for producing ultra-dense, highly aligned arrays of high-aspect ratio metal or semiconductor NWs containing up to 1400 NWs at a pitch as small as 15 nm (figure). The procedures for this SNAP method were described in the previous subchapter.

For constituting ultra-dense memory circuits whose density is far beyond what is possible with current CMOS technology, I combined these patterning methods and extremely scalable rotaxane switches, along with the defect-tolerance concepts learned from Teramac. I constructed and tested a memory circuit at extreme dimensions: the entire 160,000 bit crossbar is approximately the size of a white blood cell (~13 × 13 μm^2). At each cross-point of nanowire array, only several hundreds of rotaxanes were incorporated.

160 KBIT MOLECULAR ELECTRONIC MEMORY CIRCUITS: FABRICATION FLOW

A bottom-up approach was the key to the successful fabrication of this memory. This approach both minimized the number of processing steps following deposition of the molecular monolayer, as well as protected the molecules from remaining processing steps. In the following paragraphs, I describe the nanofabrication procedures utilized to construct the memory circuit.

Our 160,000 junction crossbar memory consists of 400 Si nanowire (NW) bottom electrodes of 16 nm width and 16.5 nm half-pitch, crossed with 400 Ti NW top electrodes of the same dimensions, and with a monolayer of bistable rotaxane molecules sandwiched in between.

My research group has previously reported on using

the SNAP technique to fabricate highly ordered arrays of 150 metal and Si NWs. For this work, the SNAP technique was extended to create 400 element NW arrays of both the bottom and top electrode materials, and so was the primary patterning method for achieving the 10^{11} cm^{-2} bit density of the crossbar.

An overview of the process flow used to fabricate the memory is shown in figure.

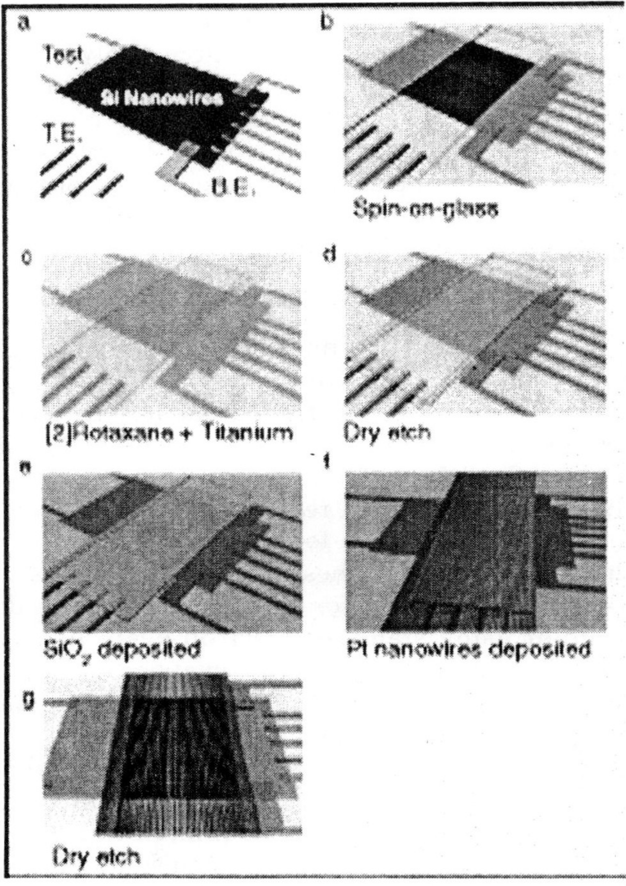

Fig. The process flow for Preparing the 160 kbit Molecular Electronic memory Circuit at 10^{11} bits/cm^2.

(a) SNAP-patterned SiNW bottom electrodes are electrically contacted to metal electrodes. (b) The entire circuit is coated with SiO_2 (using spin-on-glass (SOG)) and the active memory region is exposed using lithographic patterning followed by dry etching. (c) The bistable rotaxane Langmuir monolayer is deposited on top of the Si NWs and then protected via the deposition of a Ti layer. (d) The molecule/Ti layer is etched everywhere except for the active memory region. (e) A SiO_2 insulating layer is deposited on top of the Ti film. (f) An array of top Pt NWs is deposited at right angle to the bottom Si NWs using the SNAP method. (g) The Pt NW pattern is transferred, using dry etching, to the Ti ayer to form an array of top Ti NW electrodes, and the crossbar structure is complete.

Preparation of and contact to the bottom Si nanowire electrodes The Si NW array was fabricated as described previously.

The starting wafer for the Si NWs was a 33 nm thick phosphorous doped ($n = 5 \times 10^{19}$ cm^{-3}) silicon-on-insulator (SOI) substrate with a 250 nm thick buried oxide (Simgui, Shanghai, China). An array of Pt NWs was transferred onto this substrate using the SNAP method, and reactive ion etching was used to transfer the Pt NW pattern to form a ~2 millimeter long array of Si NWs.

The Pt NWs were then removed, and the Si NW array was sectioned into a 30 μm long region.

Electrical contacts to these bottom Si NWs, as well as contacts that are intended for the top Ti NWs were defined at this point using standard electron-beam lithography (EBL) patterning and electron-beam evaporation to produce electrodes consisting of a 15 nm Ti adhesion layer followed by a 50 nm thick Pt electrode.

Immediately prior to metal evaporation, the Si NWs were cleaned using a gentle O_2 plasma (20 standard cubic centimeters per minute (sccm), 20 milliTorr, 10 Watts, 30 seconds) followed by a 5 second dip in an NH_4F/HF solution. After lift-off, the chip was annealed at 450 °C in N_2 for 5 min to promote the formation of ohmic contacts.

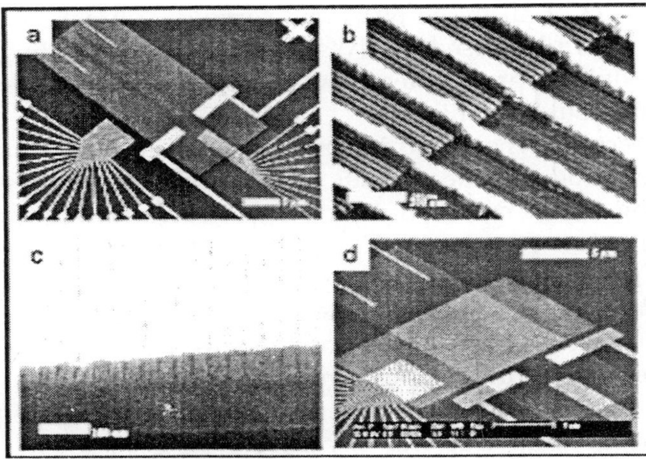

Fig.. Scanning Electron Micrographs of the Nanowire Crossbar memory Fabrication Process.

(a) A 30 micrometer long section of the SiNWs and its electrical contacts to metal leads were defined by electron-beam lithography (EBL). (b) Each electrode defined by EBL is about 70 nm wide contacting 2 ~ 4 NWs. This image illustrates that the intrinsic patterning of nanowire crossbar is beyond lithographic limits. (c) Progress-check of SOG window etching over the active memory region.

This image verifies that SOG fills the gap of NWs and SEM is a valid tool for monitoring SOG etching. (d) SOG is etched by RIE over the active memory region. Detailed processes for monitoring this etching progress are described in the text.

Figure shows an SEM image of the device at the stage in which the Si NWs and all of the external electrical contacts have been created. Note that there are four sets of EBL defined contacts. The 18 narrow contacts at the bottom left of the image will eventually connect to the top Ti NW electrodes and are used for testing of the final memory circuit. The 10 narrow contacts to the Si NWs at the bottom right of the image are also used for testing of the memory circuit. Finally there are two narrow test electrodes at the top left and two wide electrodes at the bottom right. The wide electrodes contact about 2/3 of all the Si NWs and serve dual functions. First, they ground unused Si NWs during memory testing (this procedure approximates how a fully multiplexed crossbar circuit would be utilized).

Second, when used in conjunction with the two narrow test-electrodes on the opposite side of Si NW array, they enable testing of the conductivity of the Si NWs throughout the fabrication processes. This testing procedure provided invaluable feedback for finely tuning and tracking many of the fabrication processes. Once these various contacts were established, robust Si NW conductivity was confirmed via current vs. voltage measurements. If the Si NWs were measured to be poor conductors (a very infrequent occurrence), the chip was discarded.

The device was then planarized using an optimized spin-on-glass (SOG) procedure. This planarization process was critical because the SOG not only protects Si NWs outside of the active memory region from damage that can arise during subsequent processing steps, but it also prevents evaporated Ti from entering the gaps between the Si NWs where it would be extremely difficult to remove.

Due to the extremely narrow gap between the Si NWs, this SOG step was performed in a vacuum condition: For the first generation of the devices, the SOG was spin-coated at atmospheric pressure.

However, the atmospheric spin-coating did not allow the gaps between the Si NWs to be filled completely with the SOG. The SOG penetrated only to the upper spacing of the trenches. For the complete filling, the process was done in a vacuum condition.

The substrate containing the Si NWs was placed in a small glass container covered by a rubber stopper. A needle connected to a syringe and to a diffusion pump was plugged through the rubber stopper to employ a vacuum condition. During this vacuum process, the SOG was transferred to the container via another syringe. As soon as the SOG was sucked into the container by the vacuum and therefore the substrate was covered by the SOG, the substrate was taken out immediately for a spin-coating (~ 5000 rpm, 30 sec). Before starting the planarization steps described so far, all the glasswares including the container were cleaned very carefully because even a small dust particle could ruin the device.

Especially, the top SNAP NWs process requires very clean and flat surfaces in a several millimeters range. In some cases, some dust particles appeared during the vacuum transfer despite the careful preparation.

For that case, the SOG was stripped by methanol and then the substrates were cleaned intensively by spraying methanol onto the substrates followed by blowing the dust particles off with nitrogen gas repeatedly. Upon confirming that the surface is completely dust-free, SOG was spin-coated again at an atmospheric condition. For this second SOG spin-coating, the vacuum condition was not necessary: As mentioned above, the trenches between Si NWs are not usually completely filled with the SOG if the substrate is spin-coated directly at atmospheric pressure. However, the second spin-coating performed at atmospheric pressure fills the trenches completely with the SOG, as indicated by Figure.

Next, SOG layer thinned down globally using a CF_4 plasma (20 standard cubic centimeters per minute (sccm), 10 milliTorr, 40 Watts). This etching was monitored periodically by ellipsometer and continued until the SOG layer became about 50 nm thick according to the ellipsometer. This final thickness is very critical because it affects the ensuing top SNAP and Ti layer dry-etching steps significantly. The detailed reasons are described in the paragraphs dedicated to those steps.

After globally thinning the SOG layer, an opening in photo resist was lithographically defined over the Si NWs and the tips of the 18 EBL defined contacts.

The SOG was then further etched until the tops of the underlying Si NWs were exposed. This step was monitored by periodically measuring the Si NW conductivity using the test electrodes. The majority of the dopant atoms in the Si NWs lie within the top 10 nm of the NWs. This feature means it is very straightforward to etch back the SOG without thinning the Si NWs, since the conductivity of the NWs is very sensitive to their thickness. At this stage the entire memory circuit is under SOG (and thus electrically isolated from any further top processing) except for the lithographically defined opening over the Si NWs and the 18 contacts. This opening defines the active memory region.

Deposition of Molecules and Top Electrode Materials A monolayer of bistable rotaxane switches was prepared by Langmuir-Blodgett techniques and transferred onto the device as reported previously. For the rotaxane used herein, the Langmuir-monolayers were prepared on an aqueous (18 MΩ H_2O) subphase of Langmuir-Blodgett (LB) trough (Type 611D, Nima Technology, Coventry, UK).

Before the trough was filled with the subphase, all the parts in the trough including compression barriers were cleaned very carefully by wiping with chloroform soaked wipes. Once the parts in the trough were wiped thoroughly, the filtered water was poured until the water level reached the compression barrier. From this point, the quality of the subphase was monitored by a brewster angle microscope (BAM). For further cleaning, the subphase was compressed to an area of about 50 cm^2 and then the surface of the subphase was sucked by a glass pipette connected to a pump to remove dust particles floating on the subphase surface.

As the compression and cleaning processes were repeated, the number of dust particles decreased and eventually, no dust particle was observed in the BAM image. Then, the barrier was moved back to the open position (~ 245 cm^2) and the prepared rotaxane solution was dropped onto the subphase via a syringe. The rotaxanes were prepared in a chloroform solution right before the transfer. After about 30 minutes of the chloroform evaporation, the barrier compression began at a rate of 5 cm^2/min. Once the surface pressure reached the target pressure (ð = 30 mN/m), the surface pressure was fixed and the substrate started to be pulled out at a rate of 1 mm/min. When the entire substrate was pulled out of the subphase, the step for the preparation of the Langmuir-monolayer was complete.

20 nm of Ti was then evaporated over the entire device . This Ti layer serves to protect the molecules from further top processing.

Using photolithographic techniques and BCl_3 plasma etching (10 sccm, 5 mTorr, 30 Watts), the molecule/Ti layer was then everywhere removed except for the memory active region where electrical contact to the underlying Si NWs is made. Next, a thin SiO_2 layer (~ 15 nm) was deposited over the entire substrate to isolate the EBL defined electrodes from the Pt NWs deposited in the next step.

Remember that the SOG layer was about 50 nm thick after the SOG global etching step as described in the previous paragraph so that the EBL defined electrodes as thick as 65 nm were exposed until the SiO_2 layer deposition.

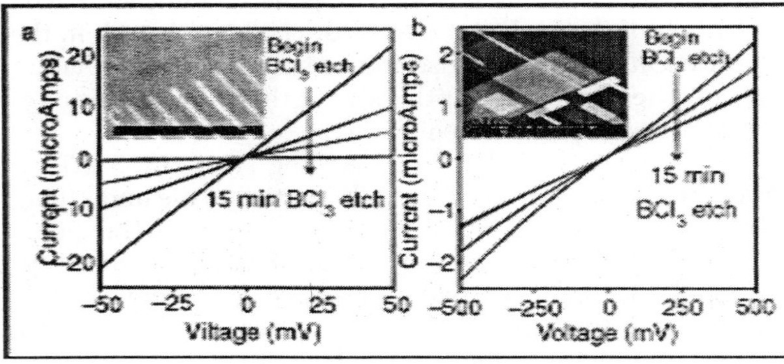

Fig. : Conductance Monitoring during the Ti layer Etching.

(a) Cross-conductance measurements between electrical contacts to the top nanowire array were performed to monitor the Ti layer etching. When the current drops to sub-10 nanoAmps, the top Ti electrodes are separated. The inset SEM image shows two representative contacts to the top Ti electrodes as highlighted in yellow. It is the cross-conductance between such contacts that was used for this measurement. (b) The conductivities of SiNWs were measured throughout the Ti layer etching to ensure that SiNWs were not damaged. The SEM image (inset) shows the current pathway that was measured. Using the SNAP technique, an array of 400 Pt NWs was then deposited over the Ti/SiO_2 layer and perpendicular to the underlying Si NWs. For the deposition of the Pt NWs, a different epoxy mixture (5 ml of THF, 5 drops of dibutylphthalate, 10 drops of epoxy and 1 drop of curing agent), compared to the one used for the Si NW generation, was used.

A larger portion of epoxy in the new mixture enabled to hold the SNAP masters more firmly while the epoxy mixture was being cured on a hot plate and the GaAs masters were being wet-etched. Especially, the usage of the new epoxy mixture was essential for the deposition of the top SNAP nanowire array because the surface of the substrate became relatively rough throughout many previous steps. For the similar reason, the new mixture was less vulnerable to undercut in the following BCl_3 plasma etching step. The prevention of the undercut was most challenging task in the project because it could arise from many factors correlating one another (recess depth, strength of cured epoxy, directionality of plasma etching etc.).

Finally, careful BCl_3 plasma etching (10 sccm, 5 mTorr, 30 W) was used to transfer the Pt NW pattern to the underlying SiO_2/Ti film, thus forming Ti NW top electrodes . The global SOG etching down to ~ 50 nm thickness was also critical for this top SNAP nanowire pattern transfer. In the devices that maintained a thick SOG layer, thus a deep recess over the active memory region, the epoxy was trapped in the recess to form its thick layer. The thick layer of epoxy was susceptible to undercut during the BCl_3 plasma etching and thus to have shorting problems in the top SNAP nanowires. This shorting problem is very fatal in a device performance because the yield of independent bits will decrease significantly. The etch endpoint was determined by monitoring the cross-conductance of the top Ti NWs. Complete transfer of the Pt NW pattern to the underlying Ti film was indicated by a fall in the cross-conductance to about 10 nS. Note that the cross-conductance does not go to zero since the Ti electrodes, while physically separated, are still electrically coupled through the crossbar junctions and the underlying Si NWs.

The health of the underlying Si NWs throughout the Ti-etching steps was also monitored as shown in figure . In most cases, the devices that skipped the SOG planarization step lost the Si NW conductance completely before the cross-conductance fell down to a value corresponding to the complete NW pattern transfer, indicating that Si NWs were damaged significantly during the BCl_3 plasma etching. Once BCl_3 plasma etching is done, the device is ready for testing. SEM images for final devices are presented in figure 3-8 at different resolution.

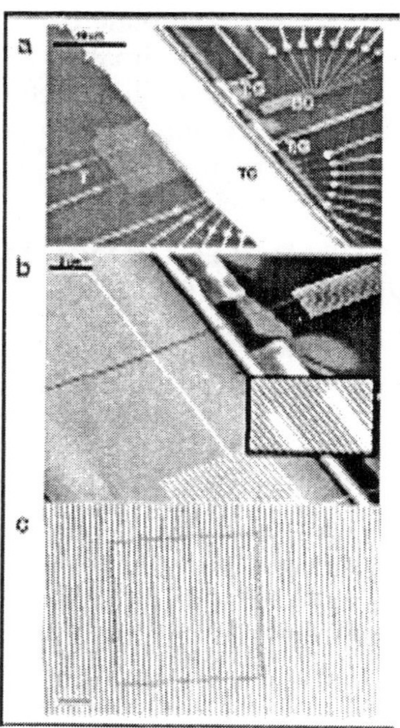

Fig. Scanning Electron Micrographs (SEMs) of the NW Crossbar Memory.

(a) Image of the entire circuit. The array of 400 bottom Si NWs is seen as the light grey rectangular patch extending diagonally up from bottom left. The top array of 400 Ti NWs is covered by the SNAP template of 400 Pt NWs, and extends diagonally down from top left. Testing contacts (T) are for monitoring the electrical properties of the Si NWs during the fabrication steps. Two of those contacts are also grounding contacts (G), and are used for grounding most of the Si NWs during the memory evaluation, writing, and reading steps. Electron beam lithography patterned 18 top (TC) and 10 bottom (BC) contacts are also visible.

The scale bar is 10 micrometers. (b) An SEM image showing the cross-point of top and bottom NW electrodes. Each cross point corresponds to an ebit in memory testing. (inset) The electron-beam-lithography defined contacts bridged 2-4 nanowires each. The scale bar is 2 micrometers. (c) High resolution SEM of approximately 2500 junctions out of a 160,000 junction nanowire crossbar circuit. The red square highlights an area of the memory that is equivalent to the number of bits that were tested. The scale bar is 200 nanometers.

160 KBIT MOLECULAR ELECTRONIC MEMORY CIRCUITS: DEVICE TESTING

The memory circuit was tested using a custom-built probe card and a Keithley 707A switching matrix for off-chip demultiplexing. Because SNAP NWs are patterned beyond the resolution of lithographic methods, each test electrode contacted between 2 and 4 NWs so that individual effective bit (ebit) contains between 4 and 16 crossbar junctions, but mostly 9 crossbar junctions.

All ebits were electrically addressed within the 2D crosspoint array by the intersection of one Si NW bottom electrode and one Ti NW top electrode. Individual molecular junctions were set to their low resistance or "1" state through the application of a positive 1.5 – 2.3 V pulse (voltages are referenced to the bottom Si NW electrode) of 0.2 s duration. A junction was set to its "0" or high resistance state through application of a –1.5 V pulse, also of 0.2 s duration.

To avoid switching an entire column or row of bits, the switching voltage was split between the two electrodes defining the ebit. Thus, to write a "1" with +2 V, a single Si NW electrode is charged to +1 V, while a single Ti NW electrode is set to –1V, and only where they cross does the junction feel the full +2 V switching voltage. Half-selected bits, that is, bits receiving only half the switching voltage, were never observed to switch. This half-select issue, though being a clear drawback of crossbar architecture, is overcome by distinctive characteristic of rotaxane: The voltages required to switch on/off MSTJs were uniform over broad junctions such that a half of the voltage did not perturb the junctions.

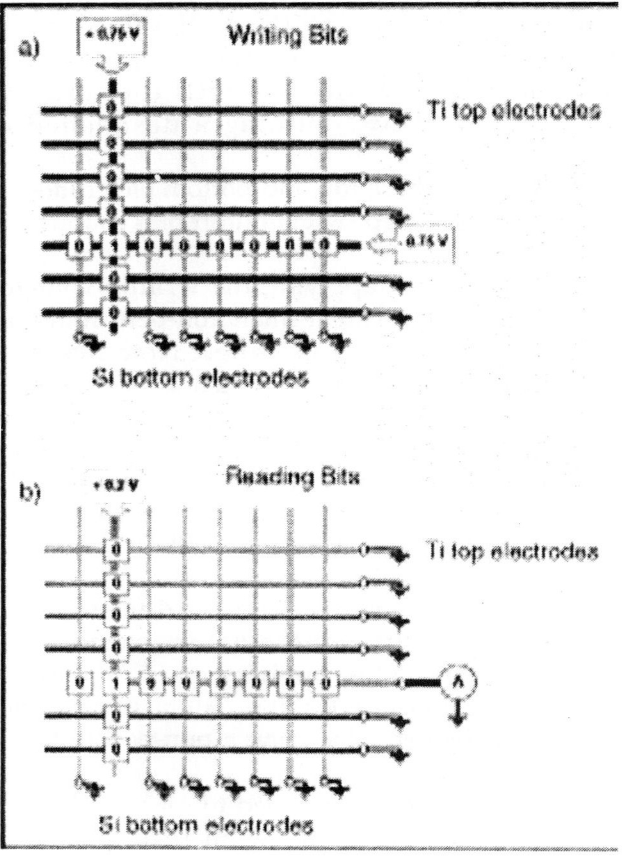

Fig. Writing and Reading Procedures in Crossbar memory Measurements. (a)

Individual ebits were read by applying a small, non-perturbing +0.2 V bias to the bottom Si NW electrode and grounding the top Ti NW electrode through a Stanford Research Systems SR-570 current pre-amplifier. Bits not being read were held at ground to reduce parasitic current through the crossbar array.

Note that all the electrical writing and reading operations described herein were done sequentially.

Schematic illustrations describing the device testing procedures composed of writing and reading bits are presented in figure Due to the half select issue, the writing bias was split into two halves of opposite polarity and each half was applied to both top and bottom electrodes, respectively, defining a designated cross-point.

Other bits along these top and bottom electrodes are not perturbed due to the sharp switch-on/off bias characteristic of rotaxanes.

(b) Before and after applying the writing voltages, the resistances of all bits are read at small non-perturbing reading bias to monitor the resistance change. Note that all other electrodes not involved in the switching of the designated cross junction stay grounded to minimize the parasitic current pathways.

By scanning electron microscopy inspection, the crossbar appeared to be structurally defect-free, with no evidence of broken, wandering, or electrically shorted NWs.

Nevertheless, electrical testing identified a large number of defective bits and the nature of those defects. This testing was done by first applying a +1.5 V pulse relative to the Si NW bottom electrodes to set all bits to '1', and then reading each ebit sequentially using a non-perturbing +0.2 V bias.

A –1.5 V pulse was then applied to set all bits to '0'. The status of each of the ebits was again read. The raw data throughout these procedures and the 1/0 current ratios are presented in figure.

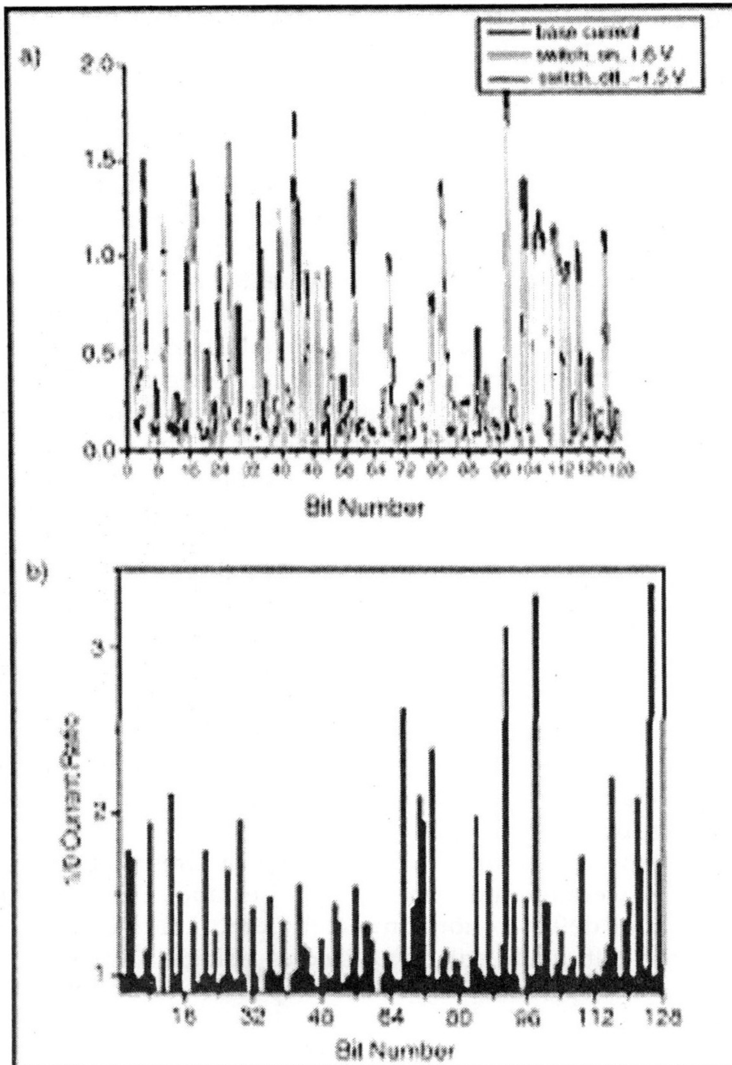

Fig. Data from Evaluating the Performance of the 128 ebits within the crossbar memory Circuit.

(a) raw current data when monitored at +0.2 V at the stage of before-switch on, after-switch-on and after-switch-off. (b)

The current ratio of the '1' state divided by the '0' state of the tested ebits. Note that many of the ebits exhibit little to no switching response. Those ebits are defective.

About 50% of the bits yielded some sort of switching response. Some of that response, however, may have originated from parasitic current pathways through the crossbar array. This is an inherent drawback of crossbar architectures wherein each junction is electrically connected to every other junction. The standard remedy is to incorporate diodes at each crosspoint, and although the molecule/Ti interface yields some rectification, we additionally grounded all NW electrodes not being used during a read or write step. By the way, the amount of rectification is dependent upon the amount of titanium oxidation that occurs at the molecule/Ti interface which, in turn, depends upon the vacuum level of the metal deposition system. For the devices reported here, the Ti was deposited at a pressure of approximately 5e-7 Torr.

For isolated devices, but constructed in a fashion similar to what was done here, this typically produces a rectification of about 10:1 at 1 V. We established a threshold for a 'good' bit based upon a minimum 1/0 current ratio of ~1.5. About 25% of the ebits passed this threshold.

Electrical testing revealed several types of defects. Bad ebits fell into a few classes, with the two most common groups being ebits that were either poor switches with little or no switching response or open circuits. Adjacent, shorted Ti top electrodes were identified when the ebits addressed by those electrodes were not independently addressable. Even though that type of defect is not completely fatal (i.e. two rows of fabricated ebits could still be utilized as a single row), we did not use ebits associated with shorted top electrode defects. The defects classified as 'switch defects' likely arose from sub-nanometer variations in the reactive ion etching process that was employed to define the top Ti crossbar NWs. Isolated devices, or crossbar memories patterned at substantially lower densities and with larger wires, can typically be prepared with a nearly 100% yield. The switch defects led to only a proportional loss in the yield of functional bits, while bad contacts or shorted nanowires removed an entire row of bits from operation. An important result from the defect map is that the good and bad bits are randomly dispersed, implying that the crossbar junctions are operationally independent of one another.

	B1	B2	B3	B4	B5	B6	B7	B8
T1	1	2	3	4	5	6	7	8
T2	9	10	11	12	13	14	15	16
T3	17	18	19	20	21	22	23	24
T4	25	26	27	28	29	30	31	
T5	33	34	35	36	37	38	39	
T6	41	42	43	44	45	46	47	
T7	49	50	51	52	53	54	55	
T8	57	58	59	60	61	62	63	
T9	65	66	67	68	69	70	71	72
T10	73	74	75	76	77	78	79	80
T11	81	82	83	84	85	86	87	88
T12	89	90	91	92	93	94	95	96
T13	97	98	99	100	101	102	103	104
T14	105	106	107	108	109	110	111	112
T15	113	114	115	116	117	118	119	120
T16	121	122	123	124	125	126	127	128

Bad SiNW contact
Good Switch
Poor Switch
Adjacent top NWs shorted (12.5%) (non-fatal)

Fig. A map of the Defective and useable ebits, Along with a pie-chart giving the Testing Statistics.

Note that, except for the bad Si NW contacts on bottom electrodes B1 and B6, and the shorted top electrodes T2 and T3, the defective and good bits are randomly distributed. Type I defects (26% of the 128 tested) are ebits that exhibited an open-circuit conductance and a low or zero amplitude switching response when tested. Type II defects (22%) are non-switchable bits that exhibited a conductance similar to that of a closed bit.

However, the ultimate test of any memory is whether it can be used to store and retrieve information. Based upon the defect map, we identified the addresses of the usable ebits, and from those addresses configured an operational memory

: the usuable bits were used to store and read out small strings of information written in standard ASCII code. The maximum number of ebits that could be tested was 180, but our electronics were configured to test 128 ebits (< 1% of the actual crossbar), and that was sufficient to demonstrate the key concepts of this memory.

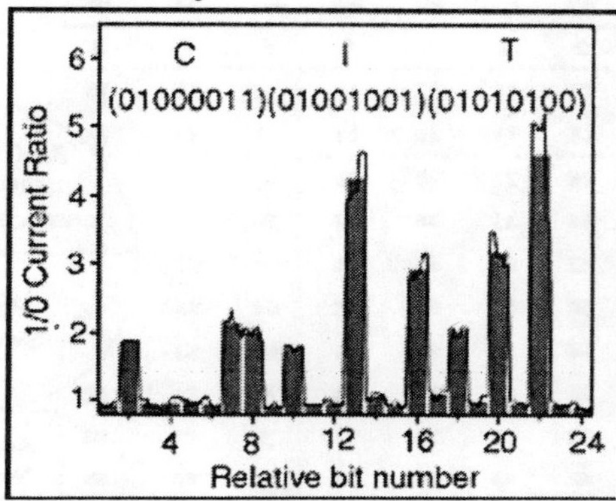

Fig. A Demonstration of point-Addressability within the crossbar.

Good ebitswere selected from the defect mapping of the tested portion of the crossbar. A string of '0's and '1's corresponding to ASCII characters for 'CIT' (abbreviation for California Institute of Technology) were stored and read out sequentially. The dotted line indicates the separation between a '0' and '1' state of the individual ebits. The black trace is raw data showing ten sequential readings of each bit while the red bars represent the average of those ten readings. Note that deviations of individual readings from their average are well separated from the threshold 1/0 line.

The solid-state switching signature of the bistable rotaxanes that were used here has been shown to originate from electrochemically addressable, molecular mechanical switching for certain device structures, but not for metal wire / molecule/metal wire junctions. In fact, our desire to utilize molecular mechanical bistable switches as the storage elements is what dictated our choice of the silicon NW / molecule / Ti NW crossbar structure. This switching signature should be effectively size-invariant, meaning that it should scale to the macromolecular dimensions of these crossbar junctions. Solid-state-based switching materials will likely not exhibit similar scaling since they arise from inherently bulk properties.

The thermodynamic and kinetic parameters describing both the bistability and switching mechanism of the rotaxane switch (and similar molecular mechanical switches have been quantified in a variety of environments. Those measurements required robust switching devices that could be cycled many times and at various temperatures.

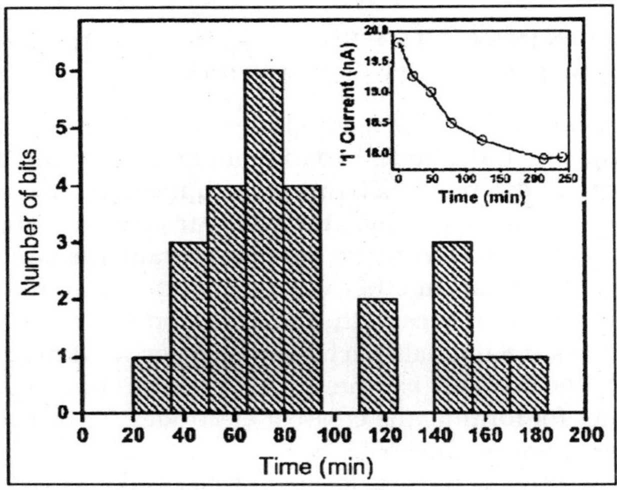

Fig.. A Histogram Representing the 1/e decay time of the '1' state to the '0' state. The 25 ebits Represented in the data were each 'large' ebits, comprised of Approximately 100 Junctions, to increase the Measurement signal to noise. Raw data from a single large ebit is shown in the inset.

The memory bits measured here were much more delicate - while all good ebits could be cycled multiple times (as evidenced by the testing and writing steps), most ebits failed after a half-dozen or so cycles, and none lasted longer than ten cycles. However, we measured the rate of relaxation from the 1Æ0 state for many of the ebits. From a device perspective, this represents the volatility, or memory retention time, of the bits. With respect to the bistable rotaxane switching cycle, this represents a measurement of the rate limiting kinetic step within the switching cycle. Our measured rate (90 ± 40 minutes; median decay = 75 minutes) was statistically equivalent to that reported for much larger (and more fully characterized) devices (58±5 minutes). Thus, our results are consistent with a molecular mechanism for the switching operation.

The volatility measurements were carried out by switching selected bits to the "1" or low resistance state, and then reading the current through those bits as a function of time.

LIMITATIONS OF THE SNAP PROCESS FOR CROSSBAR MEMORY FORMATION

The nanofabrication methods described in this chapter for creating the 160 kilobit crossbar memory circuit can be significantly extended in terms of both memory size and bit density. For our memories, the crossbar electrode materials choices have proven to be very important for successful memory operation. In other words, Si bottom electrodes and metallic top electrodes with a Ti adhesion layer were keys. Metal NWs at 8 nm half-pitch have been reported previously. Such NWs, formed by the SNAP process, only serve as templates for forming the crossbar electrodes. To be used in a crossbar memory, the SNAP NW pattern must be transferred to Si or Ti NWs for the bottom and top electrodes, respectively. Thus, it is not just the SNAP process, but the ability to translate the initially deposited SNAP NWs to form other NWs that ultimately limits the size and density of the circuitry that can be fabricated. In figure 3-3a, I present an array of 7 nm wide, 15 nm tall single crystal Si NWs patterned at 6.5 nm half-pitch. This corresponds to a crossbar that would contain approximately 6×10^{11} bits cm^{-2}. While this array may not represent the density limit of what could be achieved, densities in excess of 10^{12} cm^{-2} will be very hard to obtain using these patterning methods.

Similarly, the 160,000 junction crossbar also doesn't represent any sort of limitation. In figure 3–3b, we present SEM images of 1400 Si NWs formed using the SNAP method. Such an array size permits the formation of a 2 million bit crossbar, and it is certainly possible to further expand the concept to substantially larger structures. As mentioned, the primary limitation is that the SNAP process is that, while it is a parallel patterning method - since all nanowires within an array are created simultaneously, each array must be fabricated one at a time using a labour intensive process. A single worker, for example, can fabricate only about 20 arrays of Si NWs in a single day. However, recent advances in using nanoimprinting to replicate SNAP nanowires and to form crossbar structures indicate that high-throughput, parallel fabrication methods can be developed, even at the near molecular-densities described in this chapter.

CONCLUSION

This chapter focuses on molecular electronic memory circuits. The various generations of memory devices described in this chapter hold such common features that the devices are based on the crossbar circuit and utilize rotaxanes as information storage components. Through many generations, however, total number of bits within a single crossbar circuit and a bit density increased significantly. This scaling was possible due to the development of the fabrication procedures that allowed the integration of more delicate and higher density of electrodes with the rotaxane molecular monolayer. Despite more complicated fabrication procedures, the devices containing higher bit densities still showed the molecular switching signature. Especially, the final generation devices fabricated based upon the ultra- dense SNAP nanowire arrays also retained the molecular switching signature and exhibited a point addressability within a crossbar circuitry. Although about 75 % of the tested bits in the SNAP nanowire-based device turned out to be defective, the functional part was identified through an electrical testing and configured to write and read specific information. Notably, due to the extremely small pitch (~ 33 nm) of the SNAP nanowire array, the resultant 160 kbit crossbar memory circuits set a remarkable record in a bit density (10^{11} bits/cm^2).

Many scientific and engineering challenges, including device robustness, improved etching tools, and improved switching speed, remain to be addressed before this ultra-dense crossbar memory described here can be practical. Nevertheless, this 160,000 bit molecular memory does provide evidence that at least some of the most challenging scientific issues associated with integrating nanowires, molecular materials, and defect tolerant circuit architectures at extreme dimensions are solvable.

While it is unlikely that these digital circuits will scale to a density that is only limited by the size of the molecular switches, it should be possible to significantly increase the bit density over what is described here. Recent nanoimprinting results suggest that high-throughput manufacturing of these types of circuits may be possible. Finally, these results provide a compelling demonstration of many of the nanotechnology concepts that were introduced by the Teramac supercomputer several years ago, albeit using a circuit that contained a significantly higher fraction of defective components relative to the Teramac machine.

Chapter 14

A Fault-Tolerant Technique for Nanocomputers: Nand Multiplexing

INTRODUCTION

The emerging nanometer-scale devices will probably be less reliable than classical silicon devices for large scale integrated circuits. They are harder to fabricate and more sensitive to external influences such as electromagnetic interference, thermal fluctua-tions, radiation related effects, etc. Permanent errors may occur during manufacture while transient ones during computers' lifetime. To make systems based on nanometer-scale devices reliable, therefore, fault-tolerant architectures will be necessary.

Current fault-tolerant techniques are basically built on redundancy technologies:

- R-fold Modular Redundancy (RMR) (e.g. Triple Modular Redundancy or TMR),
- Recon¿guration.

With RMR the effect of modest transient errors are effectively eliminated, however some critical compo-nents (e.g. the Majority Voting logic in TMR) have to be highly reliable. Recon¿gurable computers have been successfully implemented in protecting against permanent errors, which are mainly generated during fabrication, while it would be much less efficient against transient ones. As the nanometer-scale devices are hopefully much smaller than current CMOS technology, the device failure rate increases due to the limit of manufacturing. In 1950s, von Neumann initiated the study of using redundant components to get reliable synthesis from unreliable components, namely, the NAND multiplexing. It was then theoretically demonstrated that with an extreme high degree of redundancy, the synthesis of unreliable components could be working reliably. In the following section 2 and 5, von Neumann's theory is brieÀy reviewed and in section 3 and 4 an alternative method is presented and some numerical evaluations are performed. In section 6 we study the error distributions in the NAND multiplexing system. Finally, there are some discussion and conclusions.

VON NEUMANN'S THEORY ON NAND MULTIPLEXING

Consider a NAND gate. Replace each input of the NAND gate as well as its output by a bundle of Q lines, as shown in Figure 1. Duplicate the NAND Q times. The rectangle U is supposed to perform a "random permutation" of the input signals in the sense that each signal from the ¿rst input bundle is randomly paired with a signal from the second input bundle to form the input pair of one of the duplicated NANDs.

Let X be the set of lines in the first input bundle being stimulated (a logic TRUE or %4%) let \ be the corresponding set for the second input bundle and let X be the corresponding set for the output bundle, as-suming the faulty probability of a NAND gate is % and the error behaves as the inversion of the output of the NAND (a von Neumann fault). Let +[X, Y, Z],have $(\overline{x}.N,\overline{y}.N,\overline{z}.N)$ elements, respectively. Clearly $(\overline{x},\overline{y},\overline{z})$, are relative levels of excitation of the two input bundles and of the output bundle, respectively.

The question is then: what is the distribution of the stochastic variable $\overline{z}$ in terms of the given $\overline{x}$ and $\overline{y}$?

With an extremely large *N*, von Neumann's theory indicates that $\overline{z}$ is a stochastic variable, approximately normally distributed, with mean $\left(1-\overline{x}\,\overline{y}) + 2\varepsilon(\overline{x}\,\overline{y} - \frac{1}{2}\right)$ and standard deviation

$$\sqrt{((1-2\varepsilon)^2\overline{x}(1-\overline{x})\overline{y}(1-\overline{y})+e(1-\overline{y})+e(1-e))/N)}$$

Therefore, given Q very large and } very small, the distribution of probability of exactly n outputs from the Q output lines of the NAND multiplexing unit
+4 {l,. 5%+{l
s
++4 5%, {+4 {,l +4
, and standard deviation
l,. %+4 %,,@Q.
being stimulated, is approximately a Poisson distribution. Empirically the Poisson Theorem gives a pretty good approximation with Q 53, } 3=38=

AN ALTERNATIVE METHOD

In this section an alternative method is given to extend the study of the NAND multiplexing technique from extreme high degree to rather low degree of redundancy, leading to a comprehensive fault-tolerant theory.

Let's consider a single NAND in the NAND multiplexing scheme, as shown in . We still as-sume that there are $\overline{x}N$ and $\overline{y}N$ input lines stimulated.

If the probabilities of error in the two input lines are independent, the probability of any output of the NAND gate that was found stimulated (by at least one nonstimulated input) is $\overline{z}'=1-\overline{x}\,\overline{y}$ (assuming that the NAND gate is faultfree). If each NAND gate has a probability ε of making a (von Neumann) error, the probability of its output being stimulated is:

$$\overline{z}=(1-\overline{x}\,\overline{y})(1-\varepsilon)+\varepsilon\overline{x}\,\overline{y}=(1-\overline{x}\,\overline{y})+2\varepsilon(\overline{x}\,\overline{y}-\frac{1}{2})$$

Thus, for each single NAND gate, the probability of the output to be stimulated (event 4) is $\overline{z}$ and the probability to be nonstimulated (event 3) is $4\,\overline{z}$. If the Q NAND gates function independently, the entire NAND multiplexing unit constitutes a Bernoulli sequence. The distribution of the probability of stimulated output lines is therefore the binomial distribution. The probability of exactly k outputs being stimulated among N output lines is then:

$$P(k) = \binom{N}{k} \bar{z}^k (1-\bar{z})^{N-k}, (k = 0,1,2,...N)$$

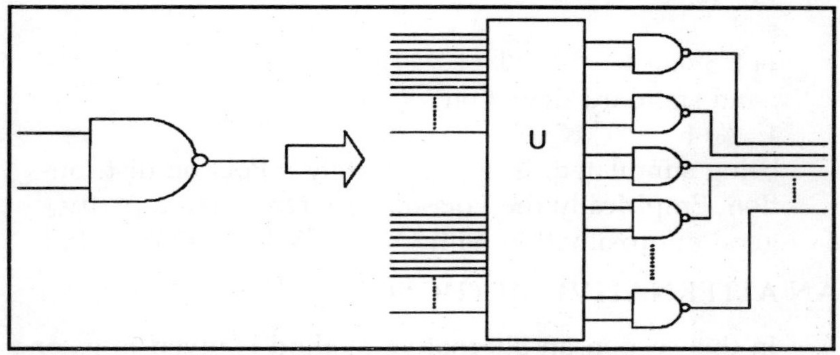

Fig. A NAND Multiplexing unit

When N is very large and $\bar{z}$ is very small, the Poisson Theorem applies. We have:

$$P(k) = \lim_{N\to\infty} \binom{N}{k} \bar{z}^k (1-\bar{z})^{N-k} = \frac{\bar{\lambda}^k e^{-\bar{\lambda}}}{k!},$$

$$(N\bar{z} = \bar{\lambda}, k = 0,1,2,...N)$$

Therefore, given N very large and $\bar{z}$ very small, the distribution of probability of exactly k outputs from the N output lines of the NAND multiplexing unit being stimulated, is approximately a Poisson distribution. Empirically the Poisson Theorem gives a pretty good approximation with $N \geq 20, \bar{z} \leq 0.05$.

To evaluate the effect of errors, the probability of possible errors below an acceptable level, i.e. $P(\bar{k} \leq x')$, needs to be computed. Since the number of the stimulated outputs is a stochastic variable, which comply with the binomial distribution, the De MoivreLaplace Theorem, when N is very large and $0 < \bar{z} < 1$, gives:

$$\lim_{N\to\infty} P\{\frac{\bar{k} - N\bar{z}}{\sqrt{N\bar{z}(1-\bar{z})}} \le x'\} = \int_{-\infty}^{x'} \frac{1}{\sqrt{2\pi}} e^{\frac{-t^2}{2}} dt$$

replacing

$$x' = \frac{x'' - N\bar{z}}{\sqrt{N\bar{z}(1-\bar{z})}}$$

then

$$P(\bar{k} \le x'')$$

$$\doteq \int_{-\infty}^{x''} \frac{1}{\sqrt{2\pi}\sqrt{N\bar{z}(1-\bar{z})}} e^{-\frac{1}{2}(\frac{t-N\bar{z}}{\sqrt{N\bar{z}(1-z)}})^2} dt$$

This shows that the probability of numbers of stimulated outputs within the N output lines of the NAND multiplexing unit $\bar{p}$ is approximately normally distributed, with mean $N\bar{z} = N((1-\bar{x}\,\bar{y}) + 2\varepsilon(\bar{x}\,\bar{y} - \frac{1}{2}))$ and standard deviation $\sqrt{N\bar{z}(1-\bar{z})} = \sqrt{((1-2\varepsilon)^2\bar{x}\,\bar{y}(1-\bar{x}\,\bar{y}) + e(1-\varepsilon))N}$, given N very large and $0 < \bar{z} < 1$.

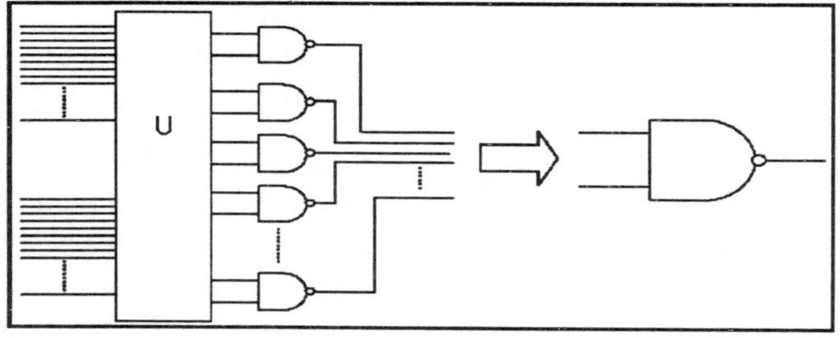

Fig. An alternative method

NUMERICAL EVALUATION

Consider next the fault distribution of the NAND multiplexing unit for different N and ε within certain ranges.

We assume that the largest possible error rate ε for a future nanoelectronic system is 3 = 4, meaning that one of ten devices is faulty on average.

Conse-quently, the ε under investigation will be in the range of. We further assume that the input excitation rates are identical to each other, i.e., $\overline{x} = = \overline{y}$. This is often true for circuits within the same device. Hence,the fault probability of the output of the NAND multiplexing unit, i.e. the probability of an output line being stimulated, becomes then:

$$\overline{z} = (1-\overline{x}^2)(1-\varepsilon)+\varepsilon\overline{x}^2 = (1-\overline{x}^2)+2\varepsilon(\overline{x}^2-\frac{1}{2})$$

For simplicity, we assume that $\overline{x}'$ of the N inputs are not stimulated, so that $\overline{x}' = 1-\overline{x}$ and $\overline{x}' < 1/2$.

Replacing $\overline{x}$ with $\overline{x}'$ in (7):

$$\overline{z} = (2\varepsilon-1)\overline{x}'^2 + 2(1-2\varepsilon)\overline{x}' + \varepsilon$$

For $\varepsilon \in [0,0.1]$, (8) is monotonically-increasing as $\overline{x}'$ varies from 0 to 1/2. For a typical $\overline{x}'$, say, 0.1, $\overline{z} \in [0.19,0.25]$. As $\overline{x}'$ increases, $\overline{z}$ will increase too. This condition does not favour a conclusion in the direction of a Poisson distribution.

We proceed with a study on the approximation of a Poisson and a Normal distribution to the binomial distribution for different sizes of the NAND multiplexing unit, i.e. for different N. We first take N = 1000.

Specifying $\overline{x}$ = 0.8 and ε = 10^{-4}, the probability (density) of the binomial, Poisson and Normal distribution are plotted in Figure. As the probability of possible errors below an acceptable level $P(\overline{k} \leq x'')$ is an important feature to evaluate the approximations, the cumulative probability distribution $P(\overline{k} \leq x'')$ for the binomial, Poisson and Normal distribution are plotted as well, in Figure.

We see that in both figures that the Normal distribution is in good accordance with the binomial distribution, while the Poisson distribution is not. The approximation for the

Normal distribution is very well kept when $\overline{x}$ varies in the range and ε varies in the range. Obviously, the larger *N* is, the better the approximation. If *N* is large enough, the error probability of the NAND multiplexing unit is approximately normally distributed.

Now consider the case that *N* = 100. We still set $\overline{x} = 0.8$ and $\varepsilon = 10^{-4}$. The fault probability and cumulative distributions are shown in Figures, respectively.

In Figure, the probability density of the Normal distribution fits in quite well with the samples of the binomial distribution. On the other hand, as the total samples *N* is not so large here, the discrete binomial distribution is no longer appropriately described by the Normal distribution.

Therefore, as shown in Figure, neither Normal nor Poisson gives good approximation to the binomial for the cumulative distribution.

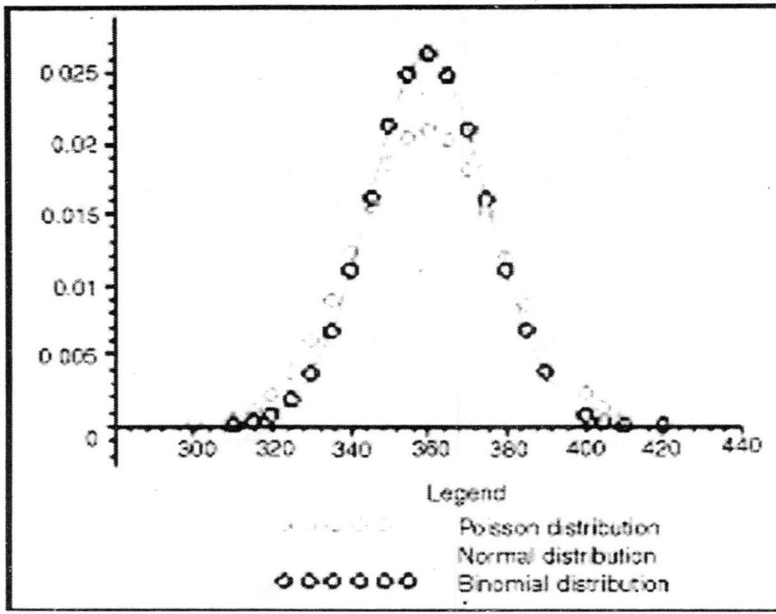

Fig. The Fault Distributions ($N = 1000$, $e = 10^{-4}$)

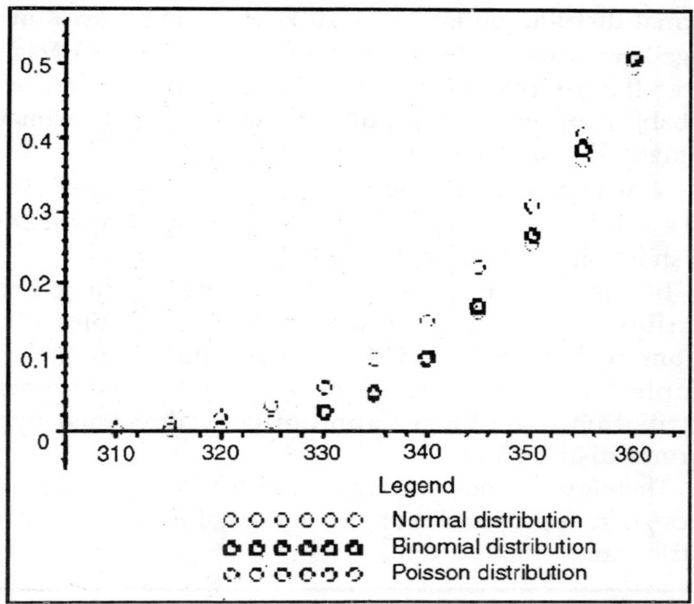

Fig. The Cumulative fault Distributions ($N = 1000$, $e = 10^{-4}$)

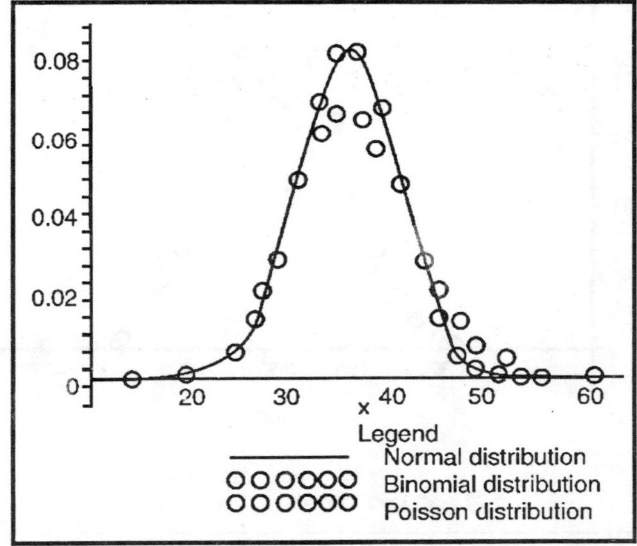

Fig. The Fault Distributions ($N = 100$, $e = 10^{-4}$)

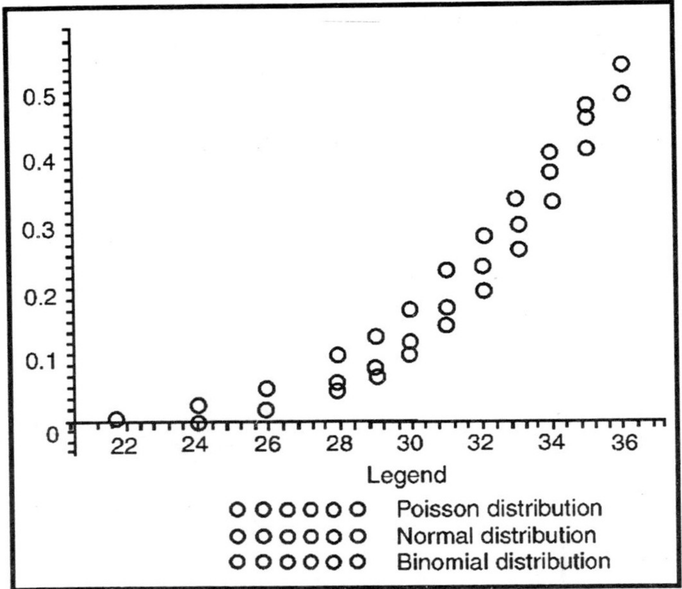

Fig. : The Cumulative fault Distributions (N = 1000, e = 10^{-4})

THE RESTORATIVE UNIT

We have discussed the setup of a NAND multiplexing unit, which executes multiple NAND operations in parallel. We considered the case where the two input bundles have almost the same stimulated levels (which is likely in circuits), i.e. $\overline{x} == \overline{y}$, with $\overline{x}, \overline{y}$ the number of stimulated lines of the two input bundles. The error behaviour of the NAND multiplexing unit can intuitively be described as follows:

- If almost all lines of one input bundle are stimulated and almost all lines of the other bundle are non-stimulated, then the probability of the out-put errors (being non stimulated) will approxi-mately be the same as the error probability in either one of the input bundles
- If almost all lines of both input bundles are non-stimulated, then the probability of the output errors

(being non stimulated) will be smaller than the error probability in either one of the input bundles

We again assume that $\overline{x}'$ is the error level of the inputs (being nonstimulated), hence, $\overline{x}' = 1 - \overline{x}$. Then the level of the stimulated outputs $\overline{z}$ is given by equation (8).

If ε is very small (8) becomes:

$$\overline{z} = 2\overline{x}' - \overline{x}'^2$$

The (9) is plotted as:

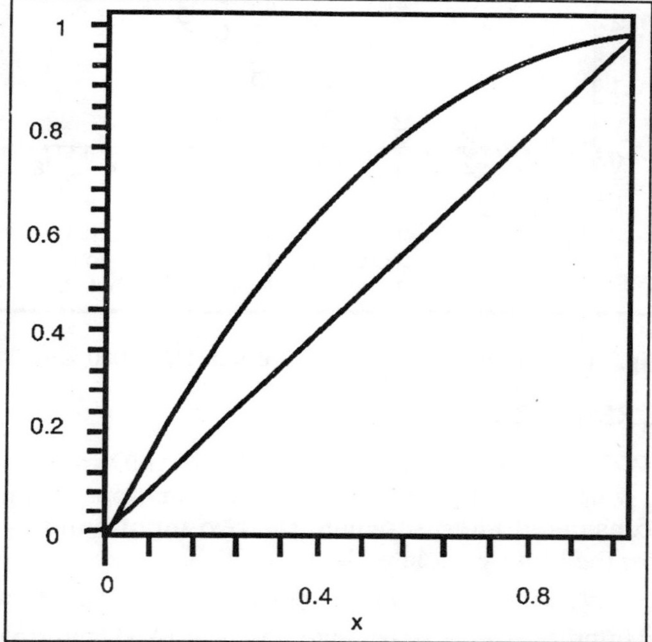

When $\overline{x}'$ varies from 0 to 1/2, $\overline{z}$ is monotonically increasing and $\overline{z} \geq \overline{x}'$. This indicates that the nonstimulated inputs give rise to more stimulated outputs, i.e. the error level is amplified. If, for example, the original error probability was 0.2, the output error probability is 0.36. Consequently, we need a unit that restores the original stimulation level without destroying the NAND function.

Let's assume a multiplexing system with two types of units. The first being the executive unit which performs the

NAND function and the second a restorative unit which annuls the degradation caused by the first unit.

The restorative unit can be made by using the same NAND multiplexing technique when duplicating the outputs of the executive unit to form the input bundles of the restorative unit, as shown in Figure.

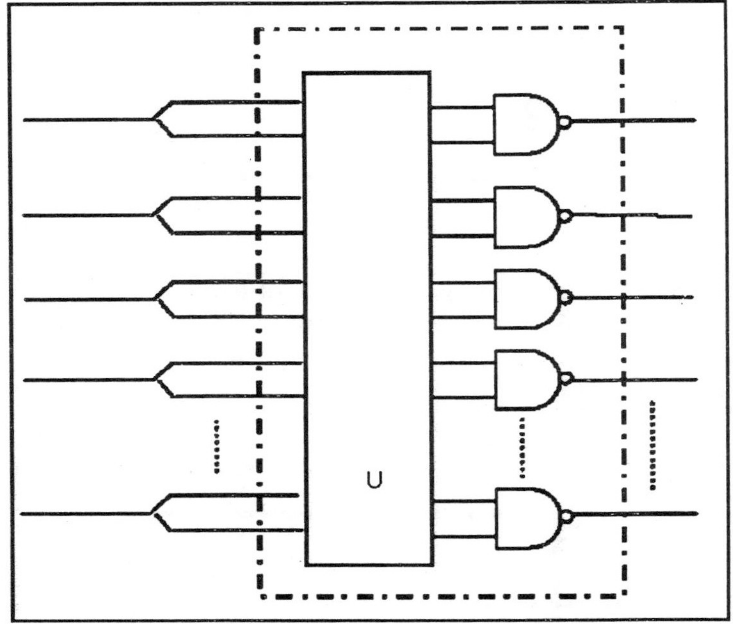

Fig. A Restorative Unit

If $\bar{x}N$ of the N incoming lines are stimulated and ε is very small, according to equation (7), the probability of any output of the restorative unit being stimulated $\bar{z}$ is approximately:

$$\bar{z} = 1 - \bar{x}^2$$

Plotting $\bar{z}$ against $\bar{x}$:

We see that instead of restoring the excitation level, the restorative unit inverts the output of the executive unit, i.e. it transforms the most stimulated bundles to most non-stimulated and vice versa. This suggests that the operation needs to be iterated to construct a proper restoration.

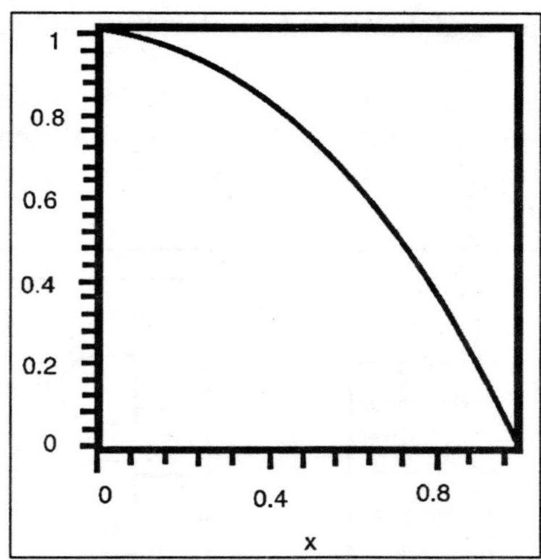

Now let the restoring unit consist of two of the restoration units in series, as shown in Figure. This unit transforms an input excitation level $\overline{x}N$ into an output excitation level of (approximately): $\overline{z}$

$=1-(1-\overline{x}^2)^2 = 2\overline{x}^2 - \overline{x}^4$

The curve of $\overline{z}$ against $\overline{x}$ is shown below, with $0 \le \overline{x} \le 1$.

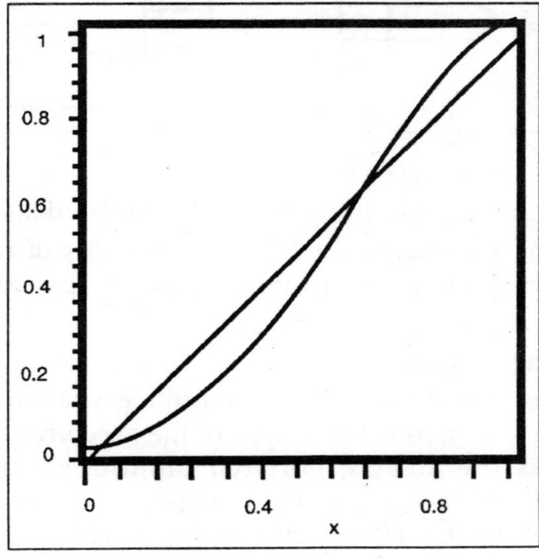

The curve intersects the diagonal $\bar{z} = \bar{x}$ three times at: $\bar{x} = 0, \bar{x}_0 = 0.618$ and $\bar{x} = 1$. If $0 < \bar{x} < \bar{x}_0$, then $0 < \bar{z} < \bar{x}$; while $\bar{x}_0 < \bar{x} < 1$ implies that $\bar{x} < \bar{z} < 1$. This indicates that the restorative unit brings every $\bar{x}$ nearer to either 0, when $\bar{x}$ is not larger than 0.618, or 1, when $\bar{x}$ is not smaller than 0.618. This process has the required restoring effect and hence the unit shown in Figure gives an effective restoration mechanism.

ERROR DISTRIBUTIONS IN A MULTI-STAGE SYSTEM

We have constructed the NAND multiplexing system of Figure. In the following parts, we study the error distributions in it.

For modest N(N < 1000)

We assume that $\bar{x}_0 N$ of the N incoming lines are stimulated for both the inputs of the executive unit in the NAND multiplexing system, and each NAND

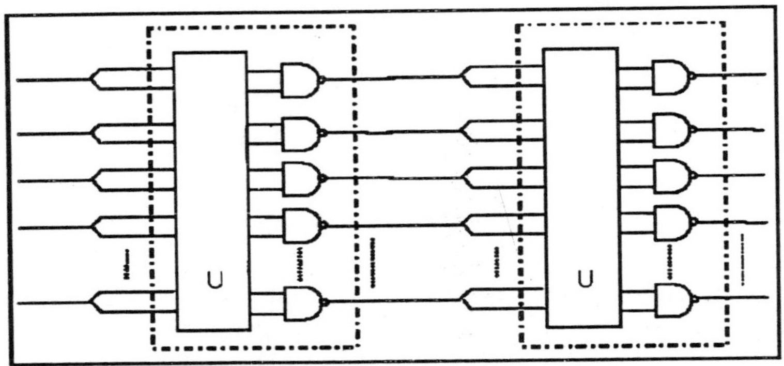

Fig. The Restorative unit

gate has a definite probability ε ofmaking a (von Neumann) error. Then the probability of any output being stimulated is:

$$\bar{z} = (1 - \bar{x}_0^2)(1 - \varepsilon) + \varepsilon \bar{x}_0^2 = (1 - e) - (1 - 2\varepsilon)\bar{x}_0^2$$

According to the results obtained in section 2, the probability distribution of the number of stimulated outputs is the binomial distribution. Hence, the probability of exactly k_1 outputs of the executive unit being stimulated in a bundle of N lines is given by:

$$P_1(k_1) = \binom{N}{k_1} \overline{z}_1^{k_1} (1 - \overline{z}_1)^{N-k_1}, k_1 \in [0,1,2,...N]$$

The outputs from the executive unit are duplicated and are connected to the successive restore unit inputs. The same multiplexing as with the executive unit is repeated and the number of stimulated inputs is $k_1 \in [0,1,2,...N]$. Hence, the stimulated input level $\overline{x}_1$ is k_1/N. The probability of any output of the first restorative unit being stimulated is then:

$$\overline{z}(k_1) = (1 - \varepsilon) - (1 - 2\varepsilon)\overline{x}_1^2 = (1 - \varepsilon) - (1 - 2\varepsilon)\left(\frac{k_1}{N}\right)^2$$

The probability of exactly k_2 outputs of the first restorative stage being stimulated under the condition of k_1 inputs is then:

$$P_2(k_2/k_1) = \binom{N}{k_2} z_2^{k_2}(k_1)(1 - \overline{z}_2)(k_1))^{N-k_2},$$

$$k_2 \in [0, 1, 2,...N]$$

Consequently, the probability of k_2 outputs being stimulated in all cases of $k_1 \in [0,1,2,...N]$ is:

$$P_2(k_2) = \sum_{k_1=0}^{k_1=N} P_2(k_2 / k_1) P_1(k_1)$$

Hereafter, the outputs from the first restorative stage are duplicated and connected to the inputs of the second restorative stage. Now the probability of Hereafter, the outputs from the first restorative stage are duplicated and connected to the inputs of the second restorative stage. Now the probability of any output being stimulated becomes:

$$\overline{z}_3(k_2) = (1 - \varepsilon) - (1 - 2\varepsilon)\overline{x}_2^2 = (1 - \varepsilon) - (1 - 2\varepsilon)\left(\frac{k_2}{N}\right)^2$$

and the probability of exactly k_3 outputs of the second

restorative stage being stimulated under the condition of k_2 inputs is:

$$P_3(k_3 / k_2) = \binom{N}{k_3} \bar{z}_3^{k_3}(k_2)(1 - \bar{z}_3)(k_2))^{N-k_3},$$

$$k_3 \in [0, 1, 2, \ldots N]$$

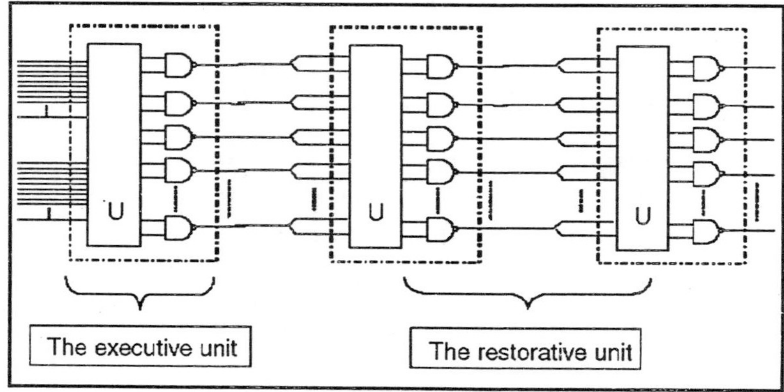

Fig. The NAND Multiplexing System

In all cases the probability of k_3 outputs of the second restorative stage, i.e. of the complete NANDmultiplexing system, being stimulated is then:

$$P_3(k_3) = \sum_{k_2=0}^{k_2=N} P_3(k_3 / k_2) P_2(k_2)$$

$$= \sum_{k_2=0}^{k_2=N} P_3(k_3 / k_2) \Big(\sum_{k_1=0}^{k_1=N} P_2(k_2 / k_1) P_1(k_1) \Big)$$

In equation the most significant parts are the conditional probabilities, $P_3(k_3/k_2)$, $P_2(k_2/k_1)$ and $P_1(k_1)$ (which can be written as $P_1(k_1/k_0)$ with fixed $k_0 = \bar{x}_0 N$.

For any identical set of inputs and outputs, all the three conditional probabilities are the binomial distribution with identical parameters, i.e.,

$$P(k/k') = \binom{N}{k_3} \bar{z}^k(k')(1 - \bar{z}(k'))^{N-k},$$

$$k, k' \in [0, 1, 2, \ldots N]$$

where

$$\overline{z}(k') = (1-\varepsilon) - (1-2\varepsilon)\left(\frac{k'}{N}\right)^2, k' \in [0,1,2,...N]$$

Therefore a $(N + 1) \times (N + 1)$ matrix Ψ, whose elements are $P(k/k'), k, k' \in [0, 1, 2,...N]$, can be made, so that all conditional probabilities for any set (k, k') are included.

$$\Psi = \begin{matrix} P(0/0) & P(0/1) & P(0/2) & .... & P(0/N) \\ P(1/0) & P(1/1) & P(1/2) & .... & P(1/N) \\ P(2/0) & P(2/1) & P(2/2) & .... & P(2/N) \\ .... & .... & .... & .... & .... \\ P(N/0) & P(N/1) & P(N/2) & .... & P(N/N) \end{matrix}$$

Accordingly, given a fixed input distribution of:

$$P_0 = [p_0, p_1, p_2 ... p_N]$$

where p_i is the probability of i inputs being stimulated, the stimulated output distribution in, and is given by:

$$P_1 = [P_1(0), P_1(1),...P_1(N)] = P_0\Psi$$
$$P_2 = [P_2(0), P_2(1),...P_2(N)] = P_0\Psi^2$$
$$P_3 = [P_3(0), P_3(1),...P_3(N)] = P_0\Psi^3$$

The output distribution of the NAND multiplexing system, therefore, can be obtained by, given the initial inputs and the conditional probability matrix.

A Stochastic Markov Chain

The number of stimulated outputs of each NAND multiplexing stage is actually a stochastic variable and its state space is $A = [0, 1, 2,...N - 1, N]$. If we name this variable $\overline{\xi}_n$ where n is the index of the individual multiplexing unit, the evolution of in the NAND multiplexing system is a stochastic process.

With fixed N and ε, the distribution of $\overline{\xi}_n$ for every n is then totally decided by the number of stimulated inputs of the nth multiplexing unit. This can be mathematically described by:

$$P(\overline{\xi}_n \in A / \overline{\xi}_1 = k_1, \overline{\xi}_2 = k_2, ...\overline{\xi}_{n-1} = k_{n-1})$$
$$= P(\overline{\xi}_n \in A / \overline{\xi}_{n-1} = k_{n-1})$$

Equation is the condition for a stochastic process to be a Markov process. The evolution of $\overline{\xi}_n$ in the NAND multiplexing system, therefore, is a Markov process, or a Markov chain for discrete states and parameters. In a stochastic Markov chain, the transition probability, which indicates the conditional probability from one specified state to another, is the most significant factor. A transition probability matrix like can be made to describe all the transition probabilities. An initial probability distribution and a transition probability matrix are then sufficient to obtain all probability distributions in such a stochastic system.

We have discussed a NAND multiplexing system with one executive and two restorative stages, where the stochastic variables can be described as $\overline{\xi}_1, \overline{\xi}_2$ and $\overline{\xi}_3$. In principle a system with any number of NAND multiplexing stages, say, $n = 5, 7, 9,...$, can be built to study the fault tolerance in it (note that the odd number is necessary to keep the NAND function). If there is a NAND multiplexing system with n individual stages in series and its transition probability matrix is given by, the output distribution of the multiplexing system is then:

$$P_n = [P_n(0), P_n(1), P_n(2), ...P_n(N)] = P_0 \Psi^n$$

In equation the transition probability matrix Ψ for each $\overline{\xi}_n$ is identical and irrelevant with regard to n. Therefore $\overline{\xi}_n$ evolves as a homogeneous Markov chain. Because none of the elements in Ψ is 0, when q gets large, approaches a constant matrix π, i.e.,

$$\lim_{n \to \infty} \Psi^n = \pi$$

and each row of π is identical with another. This, as a characteristic of a homogeneous Markov chain, indicates that not only the transition probabilities in a NAND multiplexing system will get stable and hardly changes as n becomes extremely large, but also the output distribution will become stable and independent of the stages of the multiplexing units.

N is rather large ($N > 1000$)

If N is rather large (> 1000), the output error of each NAND multiplexing stage is approximately normally distributed. If there are k stimulated inputs and corresponding k' stimulated outputs among N for each multiplexing stage, the probability density of k is given by:

$$f(k/k')$$

$$= \frac{1}{\sqrt{2\pi}\sqrt{Nz(k')(1-z(k'))}} e^{-\frac{1}{2}\left(\frac{k-Nz(k')}{\sqrt{Nz(k')(1-z(k'))}}\right)^2}$$

$$0 \le k \le N, 0 \le k' \le N$$

where

$$z(k') = (1-\varepsilon) - (1-2\varepsilon)\left(\frac{k'}{N}\right)^2, 0 \le k' \le N$$

Then the probability of each multiplexing stage having k stimulated outputs under the condition of k' inputs is approximately:

$$P(k/k') = f(k/k')\Delta k, \ldots \Delta k \sim 1$$

The probability of k outputs being stimulated in all cases for $0 \le k' \le N$ is then:

$$P(k) = \sum_{k'=0}^{k'=N} P(k/k')P(k')$$

$$= \sum_{k'=0}^{k'=N} f(k/k')P(k'))\Delta k$$

Replacing

$$P(k) = f(k)\Delta k$$

and

$$P(k') = f(k')\Delta k'$$

we then have in all cases that the probability density of k outputs being stimulated is:

$$f(k) = \sum_{k'=0}^{k'=N} f(k/k')f(k'))\Delta k'$$

In the limit we obtain:

$$f(k)=\int_0^n f(k/k')f(k'))dk'$$

The equation is an inductive expression, from which conclusions on the outputs of any NAND multiplexing system can be derived from its initial inputs.

We consider that in the multiplexing system of Figure 8, for instance, the probability density of the outputs k_3 in terms of the fixed inputs k_0 is given by:

$$f(k_3)$$

$$=\int_0^N f(k_3/k_2)(\int_0^N f(k_2/k_1)f(k_1)dk_1)dk_2$$

$$\int_0^N \int_0^N f(k_3/k_2)f(k_2/k_1)f(k_1/k_0)dk_2dk_1$$

where the initial condition is given by:

$$f(k_1)=\int_0^N f(k_1/k_0)f(k_0)dk_0=f(k_1/k_0)$$

In practice the expression is hard to be precisely computed. A practical way is to use the mean of the previous outputs as the fixed inputs of the successive stage, by which the outputs of the system in Figure 8 is distributed approximately as:

$$f(k_3)$$

$$=\frac{1}{\sqrt{2\pi}\sqrt{Nz(k_2)(1-z(k_2))}}e^{-\frac{1}{2}\left(\frac{k_3-N\,z(k_2)}{\sqrt{N\,z(k_2)(1-z(k_2))}}\right)^2},$$

$$0\le k_3\le N$$

with

$$z(k_2)=(1-\varepsilon)-(1-2\varepsilon)z(k_1)^2$$

$$z(k_1)=(1-\varepsilon)-(1-2\varepsilon)\,z_0^2$$

If, for example, a NAND multiplexing system has $N=1000$ NAND circuits and each NAND circuit has an error probability of $\varepsilon=10^{-5}$, and given that 90% of the initial inputs are stimulated, i.e., z_0 = 0.9, the outputs are approximately normally distributed, with a mean of :4 and a standard deviation of 8.

Using the normal distribution, a probability of less than 10% of outputs being faulty (stimulated) can be easily computed as being 0.9998. If the NAND multiplexing system consists of more stages of restorative units, i.e., $n \geq 5$, this probability approaches 1.

DISCUSSION

We now study the fault tolerance of a NAND multiplexing system with a fixed error rate ε for each basic logic device, while varying the I/O bundle sizes. It might be interesting to evaluate the performance of a NAND multiplexing system with $\varepsilon = 10^{-5}$ and with 90% of its inputs stimulated, and the probability that no more than 10% of its outputs is stimulated. The probability distributions versus the number of multiplexing stages are drawn in Figure for different bundle sizes $N = 10$, $N = 100$ and $N = 1000$.

If we pick the number of multiplexing stages to be n = 7, then the system has a good performance while the required redundancy (7*N*) is not too high.

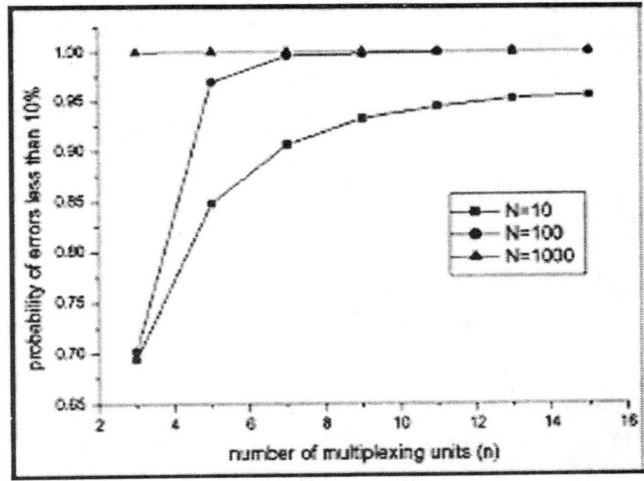

Fig. Error Distribution vs. Number of stages

The fault tolerance of this system for different error rates of the NAND circuits can be studied in this specific case. In Figure 10 the probability distribution of errors less than 10% are drawn against the error rate of individual NAND gates, with $n = 7$.

It is rather obvious that the NAND multiplexing system has a better fault tolerance with a larger bundle size N, the trade-off, however, has to be made between performance and redundancy. Another conclusion is that the NAND multiplexing technique hardly works when the error rate of the basic logic device approaches 0.1.

CONCLUSIONS

A fault-tolerant technique based on a massive duplication of imperfect devices and randomized imperfect interconnect was comprehensively studied.

Within a NAND multiplexing unit with given number of identical NAND logic gates N, input error rates $\overline{x}$ and $\overline{y}$, and error rate of the NAND logic ε, the probability of numbers of faulty outputs is theoretically a binomial distribution. It can be approximated by a Normal distribution when N is large (> 1000).

The NAND multiplexing system could have more stages of multiplexing units to improve the fault tolerance, when N is not very large.

The error distribution is then a stochastic variable, and evolves as homogeneous Markov process (chain). The characteristics of such a Markov chain are exploited. Practical simulations can be implemented to strengthen these conclusions. We search for nano-scale basic circuits and shuffle mechanisms to implement this technique.

Chapter 15

Independent Model for Nanoscale Logic Devices

INTRODUCTION

At this time, a wide variety of different mechanisms for performing digital information processing at the nanoscale have been proposed. In the literature, one finds proposals based variously on electronic, mechanical, chemical, and optical principles, and various combinations of these. Even if we narrow our attention to the all-electronic technologies, we encounter a broad range of proposed devices, based variously on semiconductors, conductors, or superconduc-tors; field-effect transistors, resonant tunneling transistors, or Josephson junctions; quantum dots or wires; metal, silicon crystals, carbon nanotubes, and organic molecules. For encoding information, electron position, voltage, current, or spin states could be used, or even atomic and nuclear configurations, motions, and spin states. This is not to mention all the possible permutations that also utilize photonic & electromagnetic phenomena, chemical transitions, etc.

We would like to provide a theoretical foundation for nanocomputing that will allow us to characterize the limits of nanocomputing, as well as to analyse, compare and optimize different candidate nanocomputer architectures. But, how are we to do this, when there is such a broad range of wildly differing technologies that have been proposed, with no clear long-term winner among them? Can we bring some order to this chaos?

One approach to this problem is to develop technology-independent theoretical models of nanocomputing, based not on the particular design constraints of any specific technology, but on more generic physical considerations that must apply to any physically possible technology.

We contend is that this technology-independent modeling effort is both feasible and useful. It is feasible because all nanotechnologies are ultimately subject to the same underlying laws of physics. At the nanoscale, the relevant "gold standard" theory is quantum electrodynamics (QED), which has stood for more than 40 years now as an extremely precise underlying model for all experimentally accessible, non-gravitational phenomena involving only photons, electrons, and stable nuclei. It thus subsumes virtually all of chemical, electrical, optical, and materials science.

Within the scope of its domain of applicability, QED's predictions have been empirically confirmed to as many as 11 decimal places of precision, and no clear contradictions between the theory and experiment have been found. For processes involving very high-energy interactions, and more exotic, unstable particles, QED has been successfully extended to yield the Standard Model of particle physics, which has reigned supreme for about 30 years now as the basis of all known physical phenomena (except gravity).

Modern theories such as QED and its relatives assure us that all physical systems and processes, regardless of their makeup, can ultimately be characterized in terms of a few universal, domain-independent physical concepts, such as entropy, energy, heat, temperature, and momentum.

Meanwhile, in computer engineering, we also ultimately care only about a range of other universal, technology-independent concepts, such as operating frequency (clock speed), energy dissipation, information propagation speed, information bandwidth and bandwidth density, heat flux, throughput, latency, performance, cost, and so forth.

In the end, any particular device technology (whether it involves carbon nanotube transistors, superconducting junctions, or spintronic valves) can be viewed as just being an interfacial "glue" layer, which executes a mapping (though possibly a complex one) between one essentially technolo-gy-independent domain (that of fundamental physics) and another one (that of computer engineering).

Thus, we ought to be able to model devices in all nanocomputing technologies generically, by abstractly characterizing how they carry out this mapping.

What are the advantages of this unified approach, as opposed to using a different, specific model for each different device technology? (One disadvantage is that a technology-specific model might well be more accurate.)

The advantages of a generic model are that:

- We are not forced to make a guess (which would prob ably be wrong anyway) concerning which specific nanocomputing technolog(y/ies) will be commercially viable 30 years from now, and thus are worth the time of developing a detailed theory for modeling them;
- The model can be easily adapted to quantitatively fit whatever specific nanocomputing technology does even-tually become dominant.
- Barring an (extremely unlikely) discovery of a huge flaw in modern fundamental physics that has eluded detection by large swarms of researchers for many decades, general qualitative results obtained from our generic model cannot become obsolete as new device technology concepts are developed. At most, certain quantitative predictions will need to be further refined.
- The model provides a framework that device physicists can use to translate from the low-level characteristics of their specific technology to system-level figures of merit (e.g., performance per unit cost) that will apply to a complete, large-scale digital system design that is based on those devices. This will help technology designers to steer their efforts towards the most useful technologies.
- The model provides a basis for nanocomputer

architecture that is by and large independent of the nanocomputing technology that is used.

The general effort to develop and explore models of computing that are based soundly on universal physical principles I call physical computing theory. In this document, we include a brief outline of a particular theoretical physical model of computing that we are currently developing, which we call CORP (Computing with Optimal, Realistic Physics). We previously described CORP in a bit more detail in. Later, we will discuss some results obtained from the CORP model.

One technique that has been useful in building CORP is to start by first reinterpreting fundamental physics itself in computational terms, which allows us to identify the key physical concepts that impact computation.

PHYSICS AS COMPUTATION

As we previously discussed in, all of the received quantum field theories, such as QED and the Standard Model, can be approximated to any desired accuracy using the Q3M (quantum 3d mesh) model, a type of parallel quantum computer consisting of a regular 3-dimensional array of cells having only a finite number of qubits per cell (representing, e.g., the number of fundamental particle quanta of each type in that cell). Each cell continuously interacts locally with its nearest neighbors, exchanging particles by means of a Hamiltonian derivable from the Schrödinger equation, while simultaneously updating its internal state according to another Hamiltonian describing the interactions between fundamental particles. Such a mesh, at a fine level of granularity, can apparently accurately capture all of Standard Model physics.

In such computational models of physics, various important physical quantities can be given precise, well-defined computational meanings. There is not space to justify and detail all of these here, so we merely summarize the most important results:

- The physical entropy in any subsystem is just the amount of incompressible (non-decomputable) information in that subsystem. Information can be effectively incompressible either when it is unknown, or when it is known but random, or even (effectively) when it is known and non-random, if its underlying pattern of order is effectively inaccessible (such as an encrypted file, when the decryption key is lost). The maximum entropy of a system is its total physical information content, the logarithm of its number of distinguishable states. The non-entropy physical information in a system can be called extropy.

- The physical energy in any subsystem is the rate of quantum physical computation in that subsystem. (This can be given a precise meaning based on the maximum rate of rotation of quantum state vectors in Hilbert space.) The rate of useful bit-operations is $R = 2E/h$ by the Margolus-Levitin theorem ; e.g. 1 eV is 484 Tbops (trillion bit-ops per second). Heat is then just the energy in that part of the information that is entropy—i.e. it is the rate at which the random bits of physical information are changing. For example, 1 BTU (British Thermal Unit) turns out to be a rate of 3×10^{36} random bit-flips per second.
- The thermodynamic temperature of a subsystem is then, roughly speaking, the heat per bit of entropy, that is, the "clock speed" for updating of random information. E.g., each degree Kelvin is a frequency $f \approx 2\text{K}/(hk \ln 2) = 28.9$ GHz of bit-updating. We can generalize temperature to generalized temperature, which is the total energy per bit of information, even for those bits that are not entropy. A system's generalized temperature (overall physical clock speed) can be higher than its thermal temperature, although non-thermal energy tends to degrade into heat, unless the system's high-energy extropic degrees of freedom are very well-isolated from parasitic interactions that will leech off their energy into entropic (thermal) degrees of freedom.

The above observations serve as a basis for our generic

technology-independent model of computation which respects all the fundamental laws of physics.

CORP DEVICE MODEL

CORP (Computing with Optimal, Realistic Physics) is the theoretical physical model of computation that we are developing. CORP's device model is essentially a "lumped element model" of the underlying computational model of physics. That is, each device is a compound subsystem that may include a large number of underlying quantum bits of state. However, as a lumped system, it still has an energy, an entropy, and thermody-

Device				
	Coding Subsystem		Non-coding Subsystem	
Logical Subsystem	Redundancy Subsystem	Structural Subsystem	Power Subsystem	Thermal Subsystem

Conceptual hierarchy of subsystems in a CORP device. If desired, a separate timing subsystem can also be split out from the coding or non-coding subsystem namic and generalized temperatures. Further, we can still conceptualize its dynamics as decomposing into Hamiltoni-ans for its self-interaction and its interactions with neigh-boring systems.

Now, not all of the degrees of freedom in a real physical device are actually used for computation. So, we break each device down into subsystems that play different roles. Figure shows the conceptual structure of a device in the CORP model. The coding subsystem is the part of the state that is varied in a controlled way to store and manipulate logical information as part of the computation of interest. We can further divide it into the logical subsystem, the bits being represented, and the redundancy subsystem, other redundant physical bits that are included for purposes such as noise immunity and error-correction.

The rest of the device's state is its non-coding subsystem. We can break this into the structural subsystem—the part of the state that must remain unchanged if the device is to continue to operate properly—the power subsystem which supplies low-entropy energy, and the thermal subsystem the part of the state that is allowed to vary ran domly and provides a pathway for removal of waste heat.

Each of these subsystems can be itself characterized by its energy, total physical information content (entropy plus extropy), and thermal and generalized temperatures. In addition, between each pair of subsystems within a device—as well as between it and its neighbors—there is an interaction energy and a generalized interaction temperature that characterize the rate at which bits of one subsystem are changed due to interactions with the other.

Ideally, all of the device's entropy is kept isolated within the thermal subsystem, although it may tend to creep into the coding subsystem (noise) or the structural subsystem (degradation) due to unwanted parasitic interactions between the thermal subsystem and the other subsystems. In general, active error correction and structural repair mechanisms (both of which are forms of refrigeration) must be used in order to keep the coding and structural subsystems clear of entropy indefinitely.

Next, we define the device's spatial geometry (region of space occupied), and identify which physical degrees of freedom within that region make up its state. This allows us to determine what assemblages of devices can exist without overlapping. Portions of the device's coding state are identified as I/O channels for communication with neighboring devices.

Finally, we summarize the device's overall computa-tional behaviour with a quantum transition function, which is a unitary map U from its input+internal logical state to its internal+output state a moment later. Such a map is necessarily invertible, so logically irreversible operations can only be implemented by extending the part of the internal state that is involved in the transformation to include not only the logical subsystem, but also the thermal subsystem. All bits discarded from the coding subsystem thus end up as entropy in the thermal subsystem.

One way to summarize the device, for technology-independent computer-engineering purposes is to specify, in addition to its transition function, also its length ℓ, area A, volume V, information capacity I, information I/O band-width B, maximum operation frequency f_{max}, standby power consumption P_{leak} and rate of standby entropy generation S_t, its energy and entropy coefficients E_f and S_f when doing reversible operations, its energy dissipation and entropy generation E_i and S_i for logically irreversible operations, and the maximum energy and entropy flows P_{max} and $S_{t,max}$ sustainable in its power and thermal subsystems.

In previous work, we have used such models to show that architectures that are predominantly logically reversible are asymptotically faster and more cost-efficient than traditional irreversible architectures, for the broadest class of applications, whenever there is a fixed limit on either total system power, or on per-area entropy flux S_{At}.

Irreversible machines have a fundamental limit on their performance within room-temperature environments of $(100W)/(k\ 300\ K \ln 2) = 3 \times 10^{22}$ bit-operations per second, per 100 Watts of power consumption. This is only about 5 orders of magnitude beyond today's technology. Reversible computing is the only possible way to exceed this limit.

Note that these results all hold true completely independently of which domain of device technology is used (electrical, optical, mechanical, chemical). This illustrates the power of technology-independent modeling.

In the next section, we illustrate how our unified physical-computational perspective can also be applied to a more technology-specific device-level problem, of estimating the minimum entropy generation per op in field-effect devices.

MINIMUM ENTROPY/OP FOR FETS

For purposes of this analysis, let a device be characterized by the following independent parameters: T_g - Average generalized temperature for operations in the entire coding subsystem, including timing signals. E_{lb} - Energy per amount of coding-state information representing one logical bit. t_{tr} - The elapsed time of one useful logical bit-opera-tion (transition between distinguishable states of a logical bit). t_d - The average time between local decoherence events for each bit within the coding subsystem. P_{lk} - Leakage power per stored logical bit. S_t - Rate of standby entropy generation per logical bit due to parasitic thermally activated transitions.

From these, we can derive the following dependent parameters: $I_{lb} = E_{lb}/T_g$ - Physical information per logical bit. The dimensionless ratio $r = I_{lb}/\text{bit}$ is called the redun dancy factor. E.g., in a voltage-coded logic circuit node, r is the number of electron states between Fermi levels at high and low voltage states. Energy per physical bit: $E_{pb} = E_{lb}/r = T_g \cdot b = kT_g \ln 2$. Rate of physical computation per logical bit: $C_{lb} = I_b \cdot \text{step}/t_{tr} = I_b \cdot (\text{op/bit})/t_{tr} = r \cdot \text{op}/t_{tr} = (E_b/T_g)\, t_{tr}(\text{op/bit})$. Rate of energy transfer (power transfer) involved in switching each bit: $P_{tr} = E_b / t_{tr}$.

We are also subject to the following constraints: $I_b \geq 1$ bit, since a bit of logical information obviously cannot be encoded in less than 1 bit of physical information. The Margolus-Levitin relation tells us that the time to change each physical bit is lower-bounded by its energy, so $t_{tr} \geq h/2E_{pb} = h/2bT_g$. (The logical bit cannot change faster than its redundant physical bits can.) Thus, for example, if the generalized temperature of the coding subsystem is only 300K, then at least 0.115 ps are required to change a bit.

In a field-effect device switched over voltage V, $P_{tr}/P_{lk} = i_{tr}V/i_{lk}V = i_{tr}/i_{lk}$, where i_{tr} and i_{lk} are the currents during desired and leakage transitions. Now, the on/off ratio $i_{tr}/i_{lk} \leq \exp(V/kT) \approx \exp(E_{lb}/rkT) = \exp[E_{lb}/(E_{lb}/T_g \cdot \text{bit})kT] = \exp[(\ln(2)k/k)(T_g/T)] = 2^c$ where $c = T_g/T$, the ratio of generalized to thermal temperature in the coding subsystem.

Decoherence will mean that that $S_t \geq I_{lb}/t_d$, and leakage will mean that $S_t \geq P_{lk}/T$. Actually we can represent the total S_t as a sum of these factors, $S_t = I_{lb}/t_d + P_{lk}/T$. Then, the total entropy generated over a bit-cycle is $\Delta S_{lbc} = S_t t_{tr} = I_{lb}(t_{tr}/t_d) + P_{lk}t_{tr}/T$. However, from earlier we have that $P_{lk} \geq P_{tr}/2^c$, so $\Delta S_{lbc} \geq I_{lb}(t_{tr}/t_d) + P_{tr}t_{tr}/2^cT$. But now $P_{tr} = E_b/t_{tr}$, so $\Delta S_{lbc} \geq I_{lb}(t_{tr}/t_d) + E_{lb}/2^cT$. Since $E_b = I_{lb}T_g$, and $T_g/T = c$, we get:

$$DS_{\text{lbc}} \geq I_{\text{lb}}\left(\frac{t_{\text{tr}}}{t_{\text{d}}}+\frac{c}{2^{c}}\right)\text{lbit.}\left(\frac{1}{q}+\frac{c}{2^{c}}\right)$$

where $q = t_{\text{d}}/t_{\text{tr}}$ is the quantum *quality factor* and c is the *coding speedup*. The value of q can also be expressed as $T_{\text{g}}/T_{\text{d}}$ where T_{d} is the *decoherence temperature,* which is the decoherence rate per bit, or in other words the interaction temperature between coding and non-coding subsystems.

Note that this expression for entropy generation can take on arbitrarily small values, but that this requires that both the $1/q$ and $c/2c$ terms be made comparably small. Fortunately, both terms can be made small simultaneously by making T_{g}— the generalized temperature of the coding system—large relative to both T and Td.

In other words, perhaps counter-intuitively, in order to minimize the entropy generated per logical bit-operation in field-effect devices, the energy per physical bit in the coding system should be made large, relative to both thermal and decoherence temperatures in the system.

Intuitively, the coding temperature Tg needs to be larger than the decoherence temperature Td so that decoherence events don't have time to happen over the course of a logic operation. Meanwhile, it also needs to be larger than the thermal temperature, in order to suppress thermally-activated leakage of electrons over the potential energy barriers set up in the field effect devices.

We argue that a closely analogous scaling analysis ought to still hold true in any digital switching technology, since this will always involve the raising and lowering of potential energy barriers between states, activated by transitions occurring between states in other similar devices.

CONCLUSIONS

The laws of physics can ultimately be understood from a computational perspective. Similarly, computation can be fundamentally described in terms of physical concepts. By using universal physical concepts such as entropy, energy, and temperature, we can compose theoretical models of nanocomputing devices in a way that is independent of the particular nanotechnology that is being used, and obtain results that will apply to all future nanotechnologies.

In this document, we briefly outlined two types of results that have already been obtained using these methods. The first was a high-level computer systems engineering analysis showing that in the long run, reversible computing, if possible, is more cost-effective than irreversible computing. The second was a demonstration that reversible computing with arbitrarily little entropy generation per operation is indeed possible in a readily generalizable model of switched FET-like devices, even when accounting for decoherence effects and thermally-activated leakage, so long as the generalized temperature in the coding system—the maximum rate of transitions per bit—can be made large relative to ambient rates of decoherence and thermal transitions. At room temperature (300 K), thermal bits change at a rate of (300 K)(1 bit)/(h/2) = 8.7 THz.

If we (reasonably) assume that decoherence temperatures are also at around this level, then present-day GHz-speed computers are still more than 3 orders of magnitude away from the point where sub-kT computing becomes manifestly possible according to this analysis. Historically, a factor of 1,000 in frequency takes only about 20 years to achieve. But, whenever we reach this point, reversible computing principles will be absolutely vital in order to make any further progress in nanocomputer power-performance beyond it, regardless of our choice of device technology.

Chapter 16

Mechanical Nanocomputer Architectures

INTRODUCTION

As Moore's Law continues to predict the trend of continually increasing densities of transistors on ever diminishing surface dimensions, components for computer architectures are rapidly approaching sizes that can be measured in nanometers, one billionth of a meter. However, as silicon transistors become measurable on the nanoscale, certain physical properties hinder their ability to function properly as they do on the macroscale. Among these well-known physical limitations are leakage, threshold voltage control, tunneling, electro-migration, high interconnect resistance, and crosstalk. These issues can restrict electrical silicon transistor functionality to the extent that future computing designs will need to consider alternative materials and architectures.

Nanoscale computer architectures introduce other challenges as well. Processors are still manufactured using lithographic techniques. Lithography involves the use of a printing press to stamp, or otherwise etch or inscribe, a design into a smooth surface. Computer circuitry is constructed in this manner, and defective units are discarded.

Thus, there is a probabilistic factor for error in the manufacturing process. As devices decrease in size, this error becomes excessively problematic. The long-term reliability of these components is also adversely affected by diminishing size. A number of solutions have been proposed for these problems; among them is the idea of Field Programmable Gate Arrays (FPGAs), which tolerate onboard errors by reconfiguring its methods of operation.

Another problem caused by reductions in size is that the interconnection wires between computational units increasingly become a dataflow bottleneck. For logical processing units, this situation is analogous to processor-memory bottlenecks.

As a result, nanoscale computers may need to become highly reconfigurable, locally-connected devices with processing units that operate exclusively on onboard memory. This architecture would drastically change the classical von-Neumann stored-programme method of performing computations. The localized nature of this architecture will shift designs from using single processors to using multiple processor units that exploit parallelism.

Architectures that could supply "a processor for every process" may be the favored designs in the future.

NANOSCALE COMPUTER ARCHITECTURES

A number of solutions for nanoscale computer architectures have been proposed, including silicon-based resonant-tunneling devices (RTDs), which consist of tunneling diodes paired with field-effect transistors (FETs).

Other proposals include the use of carbon nanotube semiconductors. Carbon nanotubes are carbon-based molecular structures that use sheets of graphite and the spherical C_{60} molecule known as the buckyball. Halves of buckyballs constitute the ends of the tube; the shaft is formed by wrapping flat granite molecules into a cylinder.

Another solution involves replacing silicon transistors with diamondoid, carbon transistors. Diamond has properties that allow it to create semi-conductors that are superior to silicon semiconductors. They allow for higher transmission speeds due to higher electron mobility, could be fit into smaller devices due to the compact tetrahedral shape of diamond, and could operate at higher temperatures and be cooled more easily than silicon transistors. They would be less expensive to manufacture due to smaller sizes and the abundance of hydrogen and carbon, and they would operate at speeds in excess of tens of gigahertz. Unfortunately, the structure of diamond is such that it cannot be created by lithographic techniques. Current diamond synthesis is performed using Chemical Vapour Deposition (CVD), which involves growing a diamondoid surface by placing the surface in an environment of highly reactive carbon molecules. Carbon bonding occurs in random locations; as a consequence, the resulting structure contains large numbers of defects. These inaccuracies make CVD insufficient to develop diamond transistors. Positional control of atoms would be much more accurate. Such engineering would require the positionally controlled removal of hydrogen atoms and deposition of carbon molecules onto the working surface.

Perhaps the most novel solution, the subject of this research, involves a new class of nano-mechanical computing devices, first proposed by Drexler in Nanosystems. The logic components of these devices are entirely mechanical in operation, reminiscent of Charles Babbage's Analytical Engine, or the ternary computing machine invented by Thomas Fowler. In these devices, logic gates and registers are constructed from a series of molecular logic rods called interlocks and are driven by kinetic forces. Such computing devices have a subtle, yet important tradeoff with more traditional electrical circuit computers. Due to the high speed of electron transfer and the kinematics of computational machinery, nano-mechanical computer devices are slower in operation than transistor-based computers. However, because the devices are atomic in scale, the difference in speed is only approximately one order of magnitude. The advantage of nano-mechanical devices over electrical devices is that nano-mechanical computing units can have drastically thicker densities. Whereas transistors must be placed on a 2-dimensional circuit board, interlock logic gates can be stacked vertically in three dimensions. These densities are estimated to be higher than silicon transistor densities by a factor greater than 10^{11}. Consequently, this architecture provides greater opportunity for a massive network of nanoscale parallel processors.

Other benefits to nano-mechanical computing architectures are that since the devices are built from atomicallyprecise molecules, the system does not wear like a comparable system on the macroscale. This feature eliminates the need for fault-tolerance. In addition, from an analytic standpoint molecular electronics are less researched than mechanical computation, thus mechanical molecular computers are more theoretically developed. Finally, mechanical computing systems have already been physically built, as in the construction of Babbage's Difference Engine, providing a foundation of conceptual feasibility for mechanical computation.

The interlocks used for logic gates in Drexler's architecture are sliding rods that have knob protrusions which slide between one another. Depending on their position, one rod may block another or allow a rod to continue sliding along its vector of movement. A logic gate has input and output interlock rods enclosed in a smooth interlock housing. The logic gates are clocked by the displacement of a driver rod which is attached to a larger "drive rod" that provides a mechanical clocking mechanism for multiple driver rods, controlled by a motor-flywheel system. The interlock driver rod is attached to the main rod by a drive spring which aids in the displacement of the main rod in the centre of the housing. A constant force reset spring rests underneath the main rod so that its position is reset when the driver rod recedes. Logic rods have probe knobs to check if a path is blocked and gate knobs to block the paths of other rods. An alignment knob is placed in the centre of the main logic rod to give the rod two absolute bounds of displacement. This alignment knob is typically placed in the centre to maximize stability. Forces from the drive and reset springs place the rod in either extreme, allowing the rod to exist in exactly two positional states. Based on the position of the input rods, movement of the output rod is either blocked or allowed to pass. The displacement of this rod indicates the output state of the gate. These interlocking rods can be combined easily to form a logic gate that computes the output of a NAND operation. This feature is fundamentally important since it is well known that any of the other logical operations can be constructed from a series of NAND operations.

The displacement sliding time of logic rods is estimated to be approximately 0.1 ns. Estimates for clock speeds rate them in excess of 1000 MIPS and running at approximately 1 GHz, which is fairly competitive compared with today's processors. The real potential behind such systems is their ability to form massively parallel processors in dense volumes: a CPU system with 10^6 logic gate "transistors" could be created within a volume of 400 nm^3.

MOLECULAR MANUFACTURING AND MECHANOSYNTHESIS

Fundamentally, the properties of a molecule depend on the arrangement of its atoms. Interestingly, this fact gives carbon a remarkable range of properties. In the form of graphite, carbon forms sheets which are strongly bonded within each sheet but loosely bonded between layers, allowing the layers to slide easily past each other. In diamond form atoms are arranged in a strongly bonded tetrahedral lattice, which is both very strong and very lightweight. It has a strength-to-weight ratio that is 50 times greater than steel, making it an ideal structure for tools. Ralph Merkle, one of the pioneering engineers of molecular manufacturing research notes that "even the ability to manufacture only the highly restricted range of structures defined by the stiff hydrocarbons would usher in a revolution in manufacturing." Future structures made from interwoven diamandoid fibers may be the strongest structures to ever exist. Since nano-mechanical computers are built from rigid components, diamondoid structures are perfectly suited to their composition.

Building molecular mechanical computers with atomic precision will require direct control of the chemical reactions that occur between atoms and molecules. The process of manufacturing machinery through such methods is known as "Molecular Manufacturing." So far, chemistry has relied on methods of controlled, probabilistic reactions resulting from collisions between masses of atoms. Currently, with very few exceptions these reactions are not directed with exact atomic precision. Still, chemists are capable of synthesizing a large number of useful compositions. We would be able to build multitudes of novel structures if we were able to specify the bonding sites of chemical reactions with accurate precision. Tools that would give us such precision would take advantage of the fundamental concepts of positional control and self assembly. Positional control refers to the idea that we could use some kind of mechanical "arm" to grasp and manipulate atoms and molecules to direct the sites of their bonding reactions. Self assembly is a concept from biology, which refers to the idea that a sufficiently sophisticated arrangement of molecules can copy itself given the appropriate resources of atoms. Implementing either of these concepts would allow us to build atomically precise molecular structures.

The bonding reactions used in such manufacturing are mechanosynthesis reactions. Mechanosynthesis is the invocation of chemical bonding caused by positionally controlled mechanical forces. Under the right conditions, applying a force between unbonded atoms may initiate a type of chemical bond called a covalent bond. Covalent bonding occurs between atoms when the atoms share electrons in their outer-most electron shell, known as valence electrons. Atoms tend to react when their free valence electrons come into contact with other free electrons.

These unbonded valence electrons are known as radicals. Atoms naturally seek states of stability, and the atom is more stable when all of its valence electrons are bonded with other atoms. The resulting bonds are strong, and they are typically stronger when compared with bonds known as ionic bonds which form from electrostatic forces. Covalent bonds between different types of atoms have varying strengths. Positional assembly involves positional control over the bonding sites of covalent reactions. As a result of kinetic forces, covalent bonding is initiated between free radicals and the atoms are combined into one molecule.

HYDROCARBON ASSEMBLERS

In the previous discussion, we noted that both positional control and self assembly are important concepts in the molecular manufacturing of nano-mechanical devices. In fact, self-assembly is not a necessary requirement, since any chemically stable molecule can be built by positionally controlled mechanosynthesis. However, molecular structures that will have practical application, including nano-mechanical computers, will be composed potentially of billions of atoms. Pragmatically, placing each of these atoms with single positional arms is infeasible. As a result, self replication is a practical requirement for our designs to scale to useful devices. If devices can replicate themselves, the construction process can be automated and take advantage of an exponential rate of growth.

Ideally, we would create a design for a "Universal Constructor" that would be capable of building large, atomically precise structures by manipulating atoms and molecules. It would be able to initiate site-specific chemical reactions that could make and break atomic bonds as desired. Such a machine would be able to build molecular structures, including copies of the machine itself. Eric Drexler's proposal for a general-purpose molecular assembler is an example of such a constructor. Most research, however, has investigated a more specific hydrocarbon assembler, whose purpose is to manufacture diamondoid structures. The assembler itself is made of hydrogen and carbon, so the self-replication requirement can be fulfilled. Many studies explore the class of hydrocarbon assemblers because it greatly simplifies the assembler design, yet the number of structures it could possibly create would be significantly useful due to the properties of diamond and graphite. The hydrocarbon assembler is a wonderful starting point, although the universal assembler could potentially synthesize materials that exceed the strength of diamond.

The hydrocarbon assembler is a nanoscale device that uses molecular arms to exert positionally controlled kinetic forces on atoms or molecules to coerce covalent bonding. These molecular arms have highly reactive tips to grab structural molecules and are controlled by an onboard molecular computer, composed of mechanical, interlocking logic units. The positioning devices manipulate feedstock molecules with a "well-defined set of chemical reactions" to build molecular objects within its internal environment, which is surrounded by a diamondoid wall to block out "contaminant" atoms and molecules that could cause uncontrolled chemical reactions. Within the internal environment, the location of every structural molecule or atom is known within an amount of positional uncertainty caused by thermal noise. Under this assumption, assemblers do not need any type of sensor to locate molecules. The assembler has the capability to reproduce itself, and during the self-replication process, the assembler expands the wall of its internal region, which will break off like a dividing cell, replacing the parent with two offspring.

The small molecular computers in each assembler are controlled by a broadcast mechanism emanating from a macroscale computer. In this "broadcast architecture," instructions from the macroscale controller are passed through acoustic waves that vibrate through a liquid environment surrounding the assemblers. The instructions are received by mechanical pressure-actuated devices in the assemblers. These pressure activated devices initiate instructions to the mechanical logic units in the assembler computer. As a result, the assembler can be reprogrammed to make different molecular structures.

SCRIPTING THE ASSEMBLER: MOLML

Molecular machinery can require billions of atoms in the final design. Unfortunately, many of the current CAD tools for molecular simulation require manual placement of atoms. Thus, there is a need for a set of "molecular compilers," which take a high level of abstraction input for molecular components and transform them into atomically positioned devices. In the same manner, manually programming every action of a broadcast computer to instruct hydrocarbon assembler activity would be overwhelmingly tedious. Thus, there is a need for a more abstract scripting language that can easily describe the desired molecular structure of the work piece. Such a language could be automatically "compiled" into the redundant individual commands sent to the assemblers.

In the spirit of a molecular compiler, the author presents the MolML language as one attempt to represent molecular data in a high level of abstraction. MolML is an Extensible Markup Language (XML) based language that is designed to factor out the redundancy in molecular structures that have large amounts of symmetry and repetition. The language is written to provide communication between a macroscale computer and a molecular assembler, facilitating the correspondence of input instructions to the assembler. The basic motivation behind use of an XML language is that the macroscale computer easily can parse or generate XML as a high-level representation for a desired nanocomputer architecture. This architecture might be designed to take advantage of the massive parallelism that is available from arrays of nanocomputer processors. A translator would take these instructions and convert them to "assembler" code, which would then be transmitted via acoustic waves to the controller computer within a hydrocarbon assembler. The assembler would then build the device. If the input instructed the assembler to build an array of nanocomputers, it might then initiate the constructed processing units to perform the computation. The output could then be sent back to the translator (using approaches discussed in Nanosystems) and received by the macroscale computer.

MolML adheres to the requirements of XML languages, including proper nesting of element tags, matching opening and closing tags, and definition of element attributes. The top-most element is the molml element, which defines a MolML document. There is only one molml element encapsulating the entire document.

The molml element has any number of molecule child

elements. These elements are block-level and may not be nested within one another. Each molecule element represents a single molecule within the assembler, and may have a single position, velocity, and acceleration element that represent the centre of the molecule, its initial velocity at time zero, and its initial acceleration at time zero, respectively. In an actual assembler, the velocity and acceleration elements may have less significance than they do for molecular simulations, but they are still included for completeness. These three elements have attributes x, y, and z, representing the xyz components for their respective 3-dimensional vectors. For the position element, the xyz coordinates are relative to the centre of the assembler, which represents the <0, 0, 0> position vector. If any of these elements are left unspecified, their values default to zero. Units of measurement are arbitrary and left to the discretion of the implementer.

To define the atoms in the molecule, the molecule element can define any number of atom elements. The atom element has a required mass element, whose value defines the mass of the atom (necessary for simulation purposes), and a required radius element, whose value defines the radius of the atom. The units of measurement are once again arbitrary. In addition, the atom element has a position element which defines the position of the atom within the molecule by its x, y, and z attributes. The position element has a type attribute, which defines how the position should be interpreted. By default, it has a value of "relative," indicating that the position values should be added to the position values of its parent element, which in this case are the values of the position element directly inside the molecule element. This behaviour can be overridden by supplying this attribute with a value of "absolute," indicating that the position should be relative to the centre of the assembler, disregarding the positioning of its parent elements.

The molecule can be declared to be fixed in one of three orientations using the fixed element. This element is useful to indicate that the molecule should be attached to a structure that has a limited range of movement. For an assembler, this instruction is very high level and its interpretation largely would be a design decision. In particular, it is convenient for simulations. The fixed element has a type attribute which indicates the range of motion allowed for the molecule; it may have a value of "absolute", "linear", or "planar." The "absolute" value indicates that the molecule has absolutely fixed positioning in relation to the assembler and cannot move. In a similar manner, "linear" indicates the range of movement for the molecule relative to the assembler is restricted to a 1-dimensional linear movement, and "planar" indicates that the molecule is restricted to 2-dimensional movement within a plane. The fixed element has two child elements, fixed-vector and fixed-plane, which define vectors for the "linear" and "planar" fixed types. The fixed-vector element has *x*, *y*, and *z* attributes that describe the linear vector defining the range of motion. For motion within a plane, recall that a plane can be completely described by a point on the plane and a vector that is normal (perpendicular) to the planar surface.

Thus, the fixed-vector element has x, y, and z attributes, which define the point on the plane, and it has attributes normalx, normaly, and normalz, which define the normal vector. Either of these child elements or both may be specified, but only the values for the fixed type specified by the type attribute are used.

An optional set of event elements describe changes in velocity and acceleration to the molecule. In an assembler, a kinetic or attractive force would correspond with the timing of these events to alter the desired vectors.

The event element has three child elements: the time element designates a value indicating when the event should occur, the velocity-change element defines a change in velocity, and the acceleration-change element defines a change in acceleration. Both the velocity-change and acceleration-change elements have a type attribute which can have a value of "relative," indicating that the change vector should be added to the current movement vector, or "absolute," indicating that the velocity or acceleration should change exactly to the given vector. If left unspecified, the type attribute defaults to value "relative." The xyz components of the change vector are specified by x, y, and z attributes. Again, the units of measurement are left to the discretion of the implementer. In the event that the molecule has a fixed type, the change in velocity or acceleration is projected onto the range of motion.

The most important elements in this language are the group and repeat elements. The group element is specified to collect atoms and subgroups into one collection. It has a position child element that adjusts the centre of the group in the same manner as the position child element for the atom element. Its purpose is so that the repeat element can be specified for the group, which will make copies of the group along a linear path. In this manner, most of the redundant positioning of atoms can be factored out of the MolML document, instead of absolutely specifying the position of every atom. In addition to atom child elements, a group element may have any number of group child elements which can be nested in a likewise manner. As a result, an infinite depth of repetition is possible, greatly simplifying the specification of 3-dimensional structures. The group specifies one repeat child element, which contains directions for how the group should be copied. This repeat element has three child elements: repeat-vector, repeat-distance, and repeat-count. The repeat-vector element specifies x, y, and z attributes which define a linear vector that is the path along which the group should be copied. To designate the amount of spacing between repetitions, the repeat-distance element encloses a value representing the distance between centers of repeated groups. Finally, the repeat-count element has an integer value defining the number of copies of this group that should be made along the repeat vector. This number indicates the number of copies that should be made in addition to the original group. A direction attribute is specified for this element, which can be given values "forward" or "backward," indicating that the copies should be made along the path of the repeat vector or against the path of the repeat vector, respectively.

The last element is used to define constant force springs which are required in Drexler's nano-mechanical designs. The spring element behaves like the group element with respect to the fact that it encapsulates a collection of groups and atoms. It cannot have a direct repeat child element. This grouping of atoms acts as the platform molecule which is attached to the spring and is subject to the constant spring force. The other atoms in the molecule element outside the spring element act as the base molecule that the spring is attached to. Several child elements define necessary information for the spring. The value of the spring-constant element defines the spring constant needed to implement Hooke's Law for simple spring motion, the spring-vector element contains x, y, and z attributes defining the vector for the linear path of spring motion, the value of the spring-length element defines the maximum extended position of the spring molecule, and finally the springposition element contains a floating point number between 0 and 1 or a percentage indicating the displacement in relation to the length of the spring at time zero.

With this definition, it is possible to define 3-dimensional molecular structures of arbitrary complexity. The following example demonstrates a complete document representing a simple cubic structure:

```
<molml> <molecule>
<position x=”10" y=”10" z=”10" />
<group>
<repeat>
<repeat-distance>80</repeat-distance>
<repeat-count dir=”backward”>1</repeat-count>
<repeat-vector x=”0" y=”0" z=”1" />
</repeat>
<group>
<repeat>
<repeat-distance>80</repeat-distance>
<repeat-count dir=”forward”>1</repeat-count>
<repeat-vector x=”0" y=”1" z=”0" />
</repeat>
<group>
<repeat>
<repeat-distance>80</repeat-distance>
```

```
<repeat-count dir="forward">1</repeat-count>
<repeat-vector x="1" y="0" z="0" />
</repeat>
<atom>
<radius>170</radius>
<mass>12</mass>
</atom>
</group>
</group>
</group>
</molecule> </molml>
```

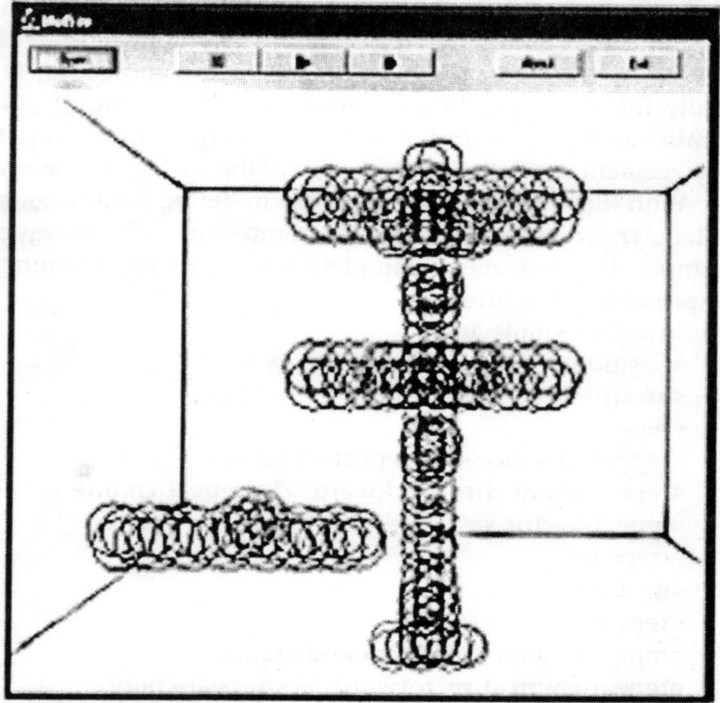

Fig. Simulation of NAND gate at Initialization

SIMULATING THE ASSEMBLED OUTPUT

To explore the feasibility of implementing the logic gates

proposed by Drexler, a rudimentary molecular dynamics simulation was written (Figure 2). The simulation itself is very simple in the sense that only simple implementations of Newtonian Laws are implemented using vector geometries, accounting for the position, velocity, and acceleration of particles in the simulation environment. The simulation uses the MolML language to define molecular structures in the environment space by parsing MolML documents to initialize the state of the simulation.

The simulation engine was tested against a MolML document that defined the structure of Drexler's nano-mechanical NAND gates. The model that was used attempted to utilize the tetrahedral lattice molecular structure of rigid diamond molecules to form the logic rods. For efficiency concerns, the housing was omitted.

The molecular design was inspired from an examination of the repetitive symmetries inherent in the lattice. For each component, only four carbon atoms needed to be declared explicitly to reproduce the diamond lattice structure by using the repeat element. This concept allowed the resulting MolML document to greatly factor out much of the redundancy in the cubic design of the logic rods. The positions of the atoms within the repetition group was computed using the carbon covalent radius of 70 pm and 109.5 degree angles between tetrahedral carbon atoms.

Using this structure, the MolML document was parsed by the simulator. The movement of the driver, input, and output rods were controlled by event elements. Under the simulation engine, the mechanics of the logic gate were successfully demonstrated for all possible inputs to the NAND operation. The driver rod was able to push back the main rod which was pushing upward by force of the reset spring. When the input rods were in place, the driver rod was retracted and the position of the main rod was determined by the state of the input rods.

CONCLUSION

As computer architectures progress into nanoscale

dimensions, the physical limitations of silicon transistors will force architects to consider innovative computing designs, a transition that is already underway.

The concept of constructing arrays of parallel processing nano-mechanical machines is particularly attractive due to their high mobility, reliability, embeddable applicability, and density which is advantageous for parallel processing networks.

Since the field of nanocomputing machinery is relatively new, much research is needed to explore their design using bottom-up methods.

At the lowest level of abstraction in this nano-mechanical computer architecture, ab initio computational chemistry and molecular dynamics simulations have demonstrated the reliability of a controlled set of covalent reactions that can be invoked by mechanosynthesis. These reactions form the groundwork for an intermediate level of abstraction, in which the family of hydrocarbon assemblers are scripted to build atomically precise molecular structures.

The structures defined by the MolML language are therefore a high-level schematic for scripting these assemblers. As a result, structures such as Drexler's interlock logic gates can be constructed from a simplified set of input data.

The next level of abstraction would investigate the interoperability of logic gates in a complete system. This concept could develop a higher-level language for constructing and placing logical units that could be translated into MolML "assembly" instructions. The design and implementation of practical nano-mechanical machinery is feasible and should be pursued.

In addition to the continued development of MolML and associated programming and simulation tools, we are actively exploring other applications of compiler design and optimization as enabling technologies, such as programme-in, chip-out (PICO), for practical use of nano-mechanical and other nanotechnology-based computer architectures. Nano-mechanical computers open an entirely new realm of computing and engineering.

The concept that many self-contained computers can exist on the nanoscale with the ability to replicate themselves introduces massive parallel processing opportunities. Such processing power opens the door for new approaches to classical computer architecture and algorithms, and presents the possibility for the invention of innovative embedded computing machines in everyday objects.

Chapter 17

Compilation for Future Nanocomputer Architectures

INTRODUCTION

Traditional compilation for high performance computers relies on a wide variety of sophisticated analyses and optimizations to translate a source language programme into efficient binary machine code for a specific target architecture. To perform this translation, the compiler parses the source programme, applying machine-independent optimizations on an abstract intermediate form of the programme. Next, machine-dependent optimizations are performed with the goal of producing hopefully a significant improvement to performance of the resulting executable programme. These improvements are made with the specific attributes of the target machine, or processor, on which the programme will run in mind.

Different processors can differ dramatically in the code improvements which work best for them. For instance, a programme compiled to run optimally on one specific architecture may perform poorly on a different architecture. As a result, at least a portion of the compiler, the back-end, must be retargeted, potentially for each different processor or class of processors. Although machine-independent optimizations can produce significant improvements in general, machine-specific knowledge, such as knowing the number of registers, memory organization and cache structure, instruction set, functional units available, and any other explicit parallelism that may be available, as with EPIC (Explicitly Parallel Instruction Computing), Superscalar and VLIW (Very Long Instruction Word) processors, is required to achieve significant results on modern, high-performance architectures.

If Moore's Law is to continue to be applicable as a predictor, significant advances in chip manufacturing techniques are needed. In modern processors, feature sizes have been reduced while heat dissipation needs have increased to the point where there is a growing consensus that Moore's Law may actually be reaching an insurmountable barrier dictated by physics this time. Nanotechnology, manufacturing performed through manipulation of atoms and molecules, is capable of overcoming this most recent barrier.

Nanocomputer architectures, produced using this molecular manufacturing approach, provide a natural successor to current general-purpose microprocessor architectures. Nanocomputers, of course, must be functionally at least as capable as their predecessors, fast, inexpensive, robust and capable of operating at room temperature and of executing legacy code. Whether Moore's Law truly loses applicability, or we merely readjust our thinking about it, it is clear that there are significant hurdles to overcome in computer design in the very near future.

Nanocomputing is discussed in this paper within the context of the topic of reconfigurable computing, which includes Field Programmable Gate Arrays (FPGA), Field-programmable Custom Computing Machines (FCCM), cellular array architectures, and synthetic neural systems, among many others. The attractive capabilities of these emerging technologies include the ability to dynamically reconfigure or redesign the functionality of the processor, to produce nanoscale features (transistors, gates, logic circuits) that are at the very limits of physics leading to processors which are many times smaller, more powerful and more efficient that those possible through current manufacturing techniques, and to generate a new, customized processor as simply as we now generate a new executable programme.

In this chapter we propose a compilation framework that could be used to produce a processor based on a source code programme such that this new processor would be ideally suited to running the compiled programme. Much as traditional compilers customize the programme to suit the machine, we propose a compiler that customizes the machine to fit the programme. Our research is on the demand side of nanocomputing rather than the supply side; although no such reconfigurable nanocomputer architecture of such a grand scale yet exists, trends point to the development of molecular- and atomic-scale switches in the near future. We present the results of preliminary experiments designed to confirm earlier results in design space exploration within a high performance EPIC machine, and extrapolate these results to the increasingly realizable promise of nanotechnology.

BACKGROUND & MOTIVATION

With limitations of miniaturization within chip manufacturing approaching the bounds of what is physically possible with known techniques, new methods of manufacturing are needed. Although some success has been achieved with extremely specialized multiprocessor systems using current technology, such as IBM's "Blue Gene" project, such large scale multiprocessor systems suffer from significant interprocessor bottlenecks. Development of scalable interprocessor communication schemes for million or billion processor systems has been slow, although it is viewed as a fundamental challenge for the future of processor design. A wavefront approach to reconfiguration and interprocessor communication for such large scale processors has been proposed as a feasible approach.

Nanotechnology provides promise as a manufacturing approach that can address these current physical limitations, and indeed research in applications of nanotechnology to logic gate design and chip manufacturing is active and well-funded in the United States, Japan and elsewhere. There is little doubt that nanocomputers, consisting perhaps of millions or billions of reconfigurable processing cells, laid out in a vast, reconfigurable fabric, are technically feasible and likely to be realized. Our work attempts to design a compilation approach that will exploit the capabilities of these highly flexible future machines.

Nanotechnology

The field of nanotechnology is a promising approach to manufacturing through the direct manipulation of atoms and molecules, accomplished through some combination of chemical, electrical and physical interactions.

Although the hype surrounding the great promise of this potentially revolutionary technique is vast, reminiscent of the ballyhoo in the 1970s regarding the coming age of artificial intelligence (a robot in every home to do one's bidding), it is undeniable that it is a viable manufacturing approach that is gaining considerable momentum.

Nanotechnology consists of a group of technologies, just emerging, which enable matter to be manipulated at the nanometer scale. This infinitesimal scale, on the order of a small number of atoms, is at the lower limits of physics as they are now understood, thus their applicability to solving current feature-size hurdles of chip manufacture, for example. Less precise applications of nanotechnology are already producing useful products in the pharmaceuticals industry, and the techniques are emerging that will allow more sophisticated products to be produced using much more precise control of matter.

Although there is controversy surrounding the future benefits of nanotechnology, current trends point to the development of artificial molecular machine systems capable of building complex systems to atomic precision. In effect, we will duplicate the ability of, for example, a cow to convert hay into beef, a feat of molecular engineering that is currently performed quite effectively by the cow. The result of this bottom-up approach to manufacturing could be products that cure cancer by repairing cells, replace fossil fuels with clean and efficient alternatives, change the colour of paint on a wall with the touch of a finger, and generate massively parallel and reconfigurable computers the size of a thumbtack. Microwave oven sized advanced molecular manufacturing systems will be capable of making large, useful products, very cheaply and with tremendous precision. Acceptance of these provocative predictions is contingent, of course, upon demonstrable results.

Along with the predictions of amazing innovations come those of enormous danger. An example of such a nanotechnological nightmare is the "gray goo" problem, where a self-replicating nanomachine, and its exponentially increasing progeny, run amuck, breaking down all of the matter encountered into a uniformly gooey gray substance.

The objective, then, is to proceed cautiously with the realization that a "bug" in a nanotechnology programme could potentially cause more damage than one that now leads to a seemingly quaint Blue Screen of Death. Exhaustive research is critical to the establishment of safe approaches to the development and use of nanotechnology.

Reconfigurable Architectures

As mentioned before, a wide variety of successful hardware approaches exist for reconfigurable computing. As physical limitations of feature size reduction and heat dissipation are reached, nanotechnology provides an approach to overcome these limitation. Because nanotechnology could lead to inexpensive production of highly reconfigurable computer hardware, it is natural to extrapolate the current state of the art to this emerging technology. Research strongly suggests that reconfigurable architectures, if efficient, will provide a better fit and thus improved performance for general purpose computation.

Recent advances in design space exploration as applied to compilation for FPGA-based and other reconfigurable architectures demonstrates the performance improvements possible with customized architectures. The concept of "programme in, chip out" (PICO) relies on compiler analysis, particularly targeted automatic parallelism, to identify programme fragments that will benefit most from customized hardware.

Future nanocomputer architectures will be formed from non-volatile reconfigurable, locally-connected hardware meshes that merge processing and memory. This architecture models more closely than any existing computer architecture that of the human brain, with its vastly superior computing power. Although the nanocomputer does not yet exist, the performance of any programme can benefit from reconfiguration of the machine on which it runs to better match the resource requirements of the programme. In contrast to an approach using design space exploration, the potential flexibility of nanocomputer architectures means that configuration will not be limited to a known design space.

Architectural Features & Compiler Analysis Techniques

Certain characteristics are common to most processors, and should be considered as candidates for reconfiguration. Characteristics such as the number of registers, functional units, memory units, cache and memory organization, branch history and prediction support, pipeline organization, and the instruction set architecture (ISA) should be considered. Compilers currently retarget a source programme to a target machine. With the flexibility of machine reconfiguration, the task of retargeting is made more powerful because the resulting target machine will be better suited to the specific characteristics and requirements of the source programme.

In analyzing the characteristics inherent in a given source programme, the ability to detect instruction-level, tasklevel and coarse-grained parallelism will also guide the description of architectural features. For instance, the amount of instruction-level parallelism (ILP) available in a programme can impact the number of functional units that are specified, leading to improved performance.

Rather than computer architects investing significant money and time in the design of an excellent general-purpose processor that does reasonably well running most source programs, the compiler itself can generate a well-tuned machine configured to match the individual source programme, allowing each to maximize performance.

The compiler has the responsibility of analyzing the source programme and generating a suitable machine description based on the results of the analysis. There are extensive analyses available, although the majority is designed to optimize the programme for a specific target machine. Any analyses that measure or enable the measurement of variable liveness, inherent parallelism, branching behaviour, common subexpressions, memory access patterns, cache utilization, register allocation opportunities, instruction scheduling opportunities, and many other characteristic-related metrics are candidates. All can provide information to guide the generation of a machine description that is suited to the source programme. The key consideration is the ability to gather information about static characteristics or predictive behaviour of the source to enable production of a best-fit machine description.

COMPILER DESIGN

The major goal of the initial work reported in this paper is the design of a compiler that generates as its output both an executable version of the original source programme and the description of a machine on which the executable will run well. This study focuses on ILP architectures, although the concepts also apply more broadly to include coarse-grain approaches. Traditional compilers take the source code and translate it into a binary form suitable for a specific processor, optimized to run as well as possible on that target machine. Knowledge of the target machine is needed to perform machine-dependent optimizations. Our approach differs in that the configuration of the target machine is unknown when compilation begins. The machine configuration is extracted from the source programme, based on its resource requirements. In this way, the resulting machine is an excellent fit to the programme.

To discover this machine description, the compiler must perform a number of analyses designed to reveal the needs of the programme. For example, live variable analysis can be performed to help determine how many registers are needed. However, the number of registers needed also depends on the issue width of the target machine, which is determined by the inherent ILP in the programme. Thus, there are numerous interactions to consider when analyzing a programme and generating a machine description.

Figure illustrates the organization of the proposed nanocompiler from a high level. Source code is processed by the Front end of the compiler, including machine-independent optimizations.

The resulting intermediate form is passed to a Machine requirements analysis phase, which performs static analysis, providing metrics to the Machine description generation phase.

The resulting machine description is used by the Processor generator phase to generate or reconfigure the target machine, and by the compiler Back end to perform machine-dependent optimizations and generate the executable code. The organization in Figure also shows how information from the Processor generator and even runtime profiling information could be fed back into the Machine requirements analysis phase to enable iterative refinement of the machine description, and thus of the processor itself.

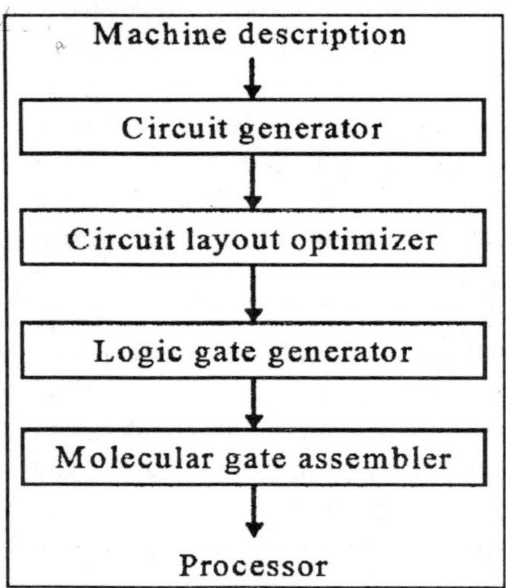

Fig. Organization of a Nanotechnology-based processor generator

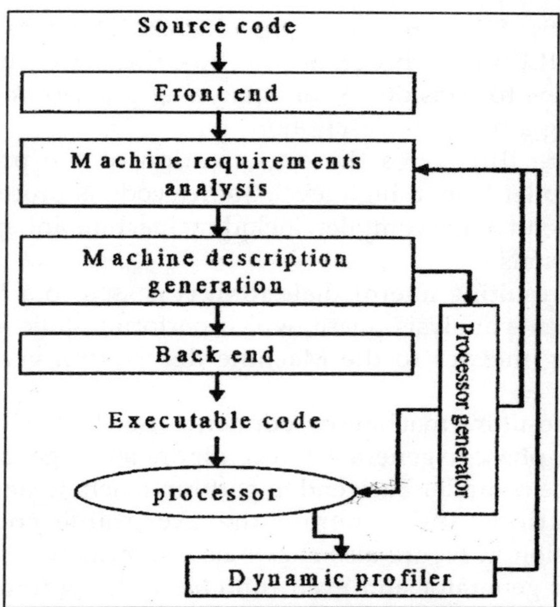

Fig. Organization of Nanocompiler

If more of the functionality of the compiler is included in the operating system or hardware, dynamic reconfiguration of the processor is possible while a programme is running. Just-in-time load-time configuration involving an analysis of the requirements of a programme could be incorporated into the processor's loader, as well. When a programme is run on a nanocomputer with a JIT reconfiguration model, the machine could be reconfigured to better suit the programme, thereby improving its performance.

Inside the Processor generator, either a machine is reconfigured or it is generated using a nanotechnology approach. Figure shows the organization of a nanotechnology-based processor generator. The machine description is analyzed through a sequence of phases that translate the description into a layout of circuits that implement the machine, which in turn is implemented using logic gates, which are assembled using molecular manufacturing techniques, to produce the target processor.

EXPERIMENTAL STUDY

The prototype of our compilation approach has been implemented as an extension of the Trimaran research compiler system. Trimaran was selected for its reconfigurable EPIC simulator and ease of extension. For this research, we modified Trimaran by isolating output from the live variable analysis and scheduling modules to incorporate their data into a machine description generator. Experiments in this initial study were facilitated with the use of a number of high-level control scripts and drivers that were developed as part of related ILP-targeted compiler optimization research.

Experiments were designed and performed to compare the impact of our nanocompiler approach on programme performance. The experiments were conducted on the full set of eight C benchmarks in the SPEC95 CINT suite.

Programs were compiled and run on five pre-configured EPIC processors and a 6th custom generated EPIC processor. These six configurations consisted of 1-4 functional units (both integer and floating point), 1-4 memory units, 1-2 branch control units (to manage branch history, prediction, etc.), and registers proportional to the number of functional units (64 per).

Note that the 6th custom generated processor varied in number of functional, memory and branch units. These simulated processors produce copious static and dynamic statistics, including machine cycles consumed, operations performed, memory and cache access behaviour, and processor utilization. The count of machine cycles was used to calculate speedup, since the simulations in Trimaran are performed independent of any specific clock cycle frequency.

We initially expected that a carefully hand-configured general-purpose processor would provide the best overall performance for the entire suite of benchmarks. The results in Table 2 indicate that reasonable and common machine configurations do indeed produce good performance speedup. Table 2 presents speedup for four machine configurations, as well as a custom, derived, machine configuration as compared with a 6th sequential, baseline processor configuration.

Table. Comparison of speedup for various architectural configurations. Note that Machine #1, a sequential processor used for baseline comparison, is not shown.

Benchmark	Machine #2	Machine #3	Machine #4	Machine #5	Machine #6: Generated machine		
(lines of source)	2F,1M,1B	2F,2M,2B	4F,2M,2B	4F,4M,4B			
	Cost: 140	Cost: 260	Cost: 280	Cost: 520	Config	Cost	Speedup
099.go (29,713)	1.55	1.67	1.71	1.71	2F,1M,2B	160	1.69
124.m88ksim (20,026)	1.89	2.31	2.33	2.33	2F,2M,1B	240	2.31
129.compress (2,322)	1.95	2.23	2.25	2.74	2F,4M,1B	440	2.71
130.li (7,683)	1.30	1.39	1.39	1.40	2F,1M,1B	140	1.39
132.ijpeg (32,214)	1.91	2.11	2.22	2.32	4F,4M,1B	460	2.32
134.perl (27,768)	1.36	1.60	1.60	1.60	2F,2M,1B	240	1.60
142.gcc (211,484)	1.61	1.89	1.89	1.89	2F,2M,1B	240	1.89
147.vortex (67,286)	1.91	2.21	2.31	2.35	2F,4M,1B	440	2.35

For each configuration, numbers of functional units, memory units and branch control units are shown. For example, "4F, 2M, 1B" signifies a machine configured with 4 functional units, 2 memory units and a single branch unit. Also shown in Table 2 is a cost for each configuration, based on a cost function C, where f, b and m are the numbers of functional, branch and memory units, respectively. The cost function is:

$$C = (fxF) + (bxB) + (mxM)$$

The values F, B and M are constant weight assigned to functional, branch and memory units to attempt to assign cost based on the complexity and increased latency of increases in the numbers of each type of unit. The values assigned to these are: F = 10, B = 20, M = 100. For example, the calculated cost of Machine #1, the sequential processor (1F, 1M, 1B), is 130.

Speedup either improved or remained steady as the cost of the simulated machine increased. Benchmarks that exhibited more memory intensive operations (129.compress, 132.ijpeg, 147.vortex) tended to continue to improve as the number of memory units increased. For benchmarks that were more computationally intensive (124.m88ksim, 132.ijpeg) or had significant data or control dependencies that constrained ILP (099.go, 134.perl, 142.gcc), additional functional or branch units tended to improve performance more than memory units.

The default configuration in the Trimaran system is Machine #3, which exhibits reasonably good speedup for all benchmarks. Results of custom generation of a machine as the result of compile-time analysis of each benchmark is shown in the last three columns of Table Note that configurations tend to reflect a reduction in the number of units when additional units did not tend to improve performance. Most notably, the cost of each generated machine is less than the cost for comparable speedup provided by one of the preconfigured machines, and the Machine requirements phase appears to have frequently identified a best, or at least better, fit machine configuration for each benchmark.

Within the context of the EPIC architectures possible in Trimaran, it appears that our compiler framework leads to reduced cost, improved speedup, or a balance between the two when there are competing factors. In some cases, four memory units did not lead to speedup over a configuration with two memory units, and the machine requirements phases arrived at the correct conclusion. Within the relatively constrained domain of EPIC, VLIW, Superscalar and similar ILP architectures, this approach may well be a feasible addition to current compilers that target reconfigurable versions of these machines.

Because a hypothetical nanocomputer architecture is quite large (billions of processing units on a single chip), the machine description generation task is more difficult. With increased flexibility and freedom of reconfiguration comes increased responsibility on the part of the compiler phases to thoroughly analyse the source programme and produce a well-fitting machine description. There are improved opportunities for parallelism, and the potential to redesign the instruction set, pipeline organization, and any other feature of the machine. Ultimately, the promise of this computational nanotechnology is straightforward, on-demand molecular-level manufacturing, which means that a new processor could be specified, manufactured, used and discarded (or recycled). Although such capabilities sound far-reaching and futuristic, from a practical standpoint such a technological revolution is technically and physically feasible, if far-off.

RELATED WORK

Research in the area of FCCMs where a system combines reconfigurable logic with a general-purpose processor has demonstrated speedup. This research has focused on architectural details of the reconfigurable hardware and compiler features that exploit it. Our research attempts to build on this work by increasing the capability of the compiler to generate more of the target machine.

Integration of a reconfigurable functional unit into a superscalar RISC processor improves upon earlier FCCM research by focusing on the hardware-compiler interface, although it relies on hand modification of source programs to achieve speedup. With our approach, all analysis and modification would be performed by the compiler.

An innovative approach to large-scale, homogeneous, undifferentiated, reconfigurable architecture improves upon FPGAs using a less expensive nanoscale "cell matrix" approach. This work describes how networks of atomicscale switches can be configured in parallel and used to fabricate scalable processors that are customized to specific tasks. Our research is targeted to a nanocomputer cell matrix architecture, focusing on use of the compiler to automatically generate reconfiguration instructions.

Hewlett Packard's research on PICO, a system that automatically designs custom computers is very similar in its goals. With PICO, parallel portions of a programme run on reconfigured FPGA hardware while the remainder runs on a connected, embedded processor. PICO employs design space exploration, selecting the best configuration from among a set of pre-designed configurations. While PICO targets primarily embedded processors and uses design space exploration, our approach envisions a desktop or embedded computer that reconfigures its own hardware to any arbitrary configuration, using machine requirements analysis and nanotechnology. This process may resemble FPGA reconfigurability, or actual physical molecular reassembly, performed at compile-time or run-time.

The field of nanotechnology is quite active and rapidly advancing, with frequent accomplishments being made in many disciplines including medicine, pharmaceuticals, chemistry, physics, and of course, computer engineering. Little has been done in the area of compilers and optimization specifically aimed at using this technology.

Our research attempts to pursue basic and applied nancompiler research and encourage others to do the same.

SUMMARY

Reconfigurable computing is an emerging area of hardware, and the promise of nanotechnology to hurdle looming barriers to continued applicability of Moore's Law is being recognized. We have designed a compiler framework targeted to a nanocomputer, and demonstrated that the techniques of machine description analysis and processor configuration-code generation using our compiler can improve performance while limiting costs. Although our implementation was restricted to a reconfigurable EPIC architecture simulator, as it was the most appropriate and available platform at the time of our study, there are similarities in the requirements of all processors, including nanocomputers. With the significant flexibility and capability of nanocomputers, it seems that the responsibility for guiding the configuration will fall to the compiler. This research demonstrates the feasibility of that approach. Where once the compiler customized the programme to suit the machine, the compiler will soon fully customize the machine to suit the programme, extending code generation to include code that will reconfigure, or even guide the manufacture of the processor itself.

We are planning extensive research and experimentation in the area of compiler optimization for nanocomputers. We are currently developing a compiler analysis framework that simulates generated architecture specifications using Colored Petri Nets. Additional work is being done in compiler approaches to generating and simulating nanomechanical computational machine descriptions from source code, applications to low-power and embedded computing, automatic parallelization techniques and description of nanocomputer architectures.

Index

D

E

F

G

H

I

L

M

N

O

P

Q

R

S

T